SPIRITS AND POWER

An Analysis of Shona Cosmology

For

Bishops Dr Rudolf Graber of Regensburg

and Johannes Lueck, formerly of Aliwal North.

SPIRITS AND POWER

An Analysis of Shona Cosmology

HUBERT BUCHER

1980
OXFORD UNIVERSITY PRESS
Cape Town

Oxford University Press

OXFORD LONDON GLASGOW
NEW YORK TORONTO MELBOURNE WELLINGTON
NAIROBI DAR ES SALAAM CAPE TOWN
KUALA LUMPUR SINGAPORE HONG KONG TOKYO
DELHI BOMBAY CALCUTTA MADRAS KARACHI

ISBN 0 19 570176 3

A generous grant of *Missio München* towards the publication of this book is gratefully acknowledged by the author.

Set in 10 on 12 pt Times
Printed in South Africa by Citadel Press, Lansdowne, Cape
Published by Oxford University Press, Harrington House,
Barrack Street, Cape Town 8001, South Africa

TABLE OF CONTENTS

LIST OF MAPS

PREFACE

This book is based on a study which I made during a year's sabbatical leave from the missions in South Africa. I had first worked here for ten years among the Xhosa and, during a further period of nearly four years, came into contact with most of the other Bantu-speaking groups in the country. None of these groups' religious assumptions seem to have received as much attention from social anthropologists as have those of the Shona in Rhodesia. The study was carried out in the hope that it would help me to contribute eventually to the better understanding of African cosmologies in South Africa itself.

Concentrating on the Shona rather than on one of the South African groups has enabled me to use a relatively large body of reliable anthropological literature. There is a further good reason for having chosen the Shona as the subject of this study. Whereas there are some two thousand Independent Churches in South Africa, there exist as yet only a comparatively small number of such Churches among the Shona, and these have been quite thoroughly studied. I have concentrated on only one branch, the 'Churches of the Spirit', whose founders received their inspiration through contact with similar Churches in South Africa.

This study addresses itself first and foremost to men and women who are engaged in the work of evangelization. However, it may be found that it is also a useful contribution to the anthropological literature on the Shona. For although it is based on secondary literature, I believe I have presented the material in a manner which throws new light on a number of important aspects of traditional Shona religion.

I wish to express my sincere thanks to Dr Wendy Johnson, under whose supervision I did this research in Oxford, for her thoughtful criticism and patient advice. Special thanks are due to Dr M. F. C. Bourdillon who gave me valuable insights into aspects of traditional Shona religion. Later both he and Dr Aylward Shorter WF read the manuscript and provided helpful criticism. The chief librarian and staff of Rhodes House Library in Oxford have been most helpful throughout my research. I am particularly grateful to Dr Rudolf Graber, Bishop of Regensburg, my home town, for having granted me leave of absence and financial support to carry out this research during 1975/76.

H.B.

FOREWORD

It was welcome news that Hubert Bucher, who has given so much time and labour to the study of African peoples, was given an episcopal see in South Africa. Time spent by a member of one race in trying to understand members of other races is time well spent in this plural society, and is rightly recognized as such by the Church. It is equally cheering to see that Bishop Bucher's episcopal responsibilities have not prevented his work in this field from coming to fruition. Indeed, a serious exploration of the meeting between Christianity and African cosmologies must remain one of the bishop's principal responsibilities.

We now have at our disposal numerous works on Shona peoples and their various institutions. There are published studies on chiefs and politics, on Shona agriculture, on their kinship and marriage laws, on Shona in towns and cities. Their traditional religion has been well covered by a number of authors, and there is now an extensive literature on new Shona religions. But there has been little attempt to explicate the central ideas which govern Shona thought and culture.

This is Bishop Bucher's welcome contribution. His study is based on a mass of published literature rather than on the detailed first-hand knowledge of some small community. The general nature of the study helps the author to detach himself from any particular community and its problems, and to see instead strands which are common to all the Shona and to a variety of institutions.

He develops the ideas of power and powers, which help him to bring together disparate aspects of Shona culture. Concern about power is shown to be the fundamental feature of Shona cosmology.

Although this analysis is in itself a worthy contribution to our understanding of the Shona, the study contains a further theme which is of broader significance for those of us who are concerned with the impact of Christianity on Africa. There has been considerable debate on whether the new Independent Churches in Africa represent a step from Christianity back to paganism, or a step away from paganism towards Christianity, with the latter view now prevaling. The prior and more fundamental question on the extent to which these Churches represent, from the cosmological point of view, any fundamental movement at all, has received little direct attention. It is clear that the Churches are growing, and that they represent and embody changes in ritual and in community allegiances: cultural changes which can be categorized broadly as socio-political. But to what extent do Independent Churches represent or embody changes in the way people think about and understand the world they live in? Bishop Bucher addresses himself to this question, and argues that there is little evidence of cosmological change as Shona leave their traditional religious practices and join the Independent Churches.

This book is about Shona cosmology. The author suggests that for the full impact of Christianity to be enjoyed by African peoples, there must be a confrontation between Christianity and the fundamental features of traditional cosmologies. That the Independent Churches have failed to produce this confrontation is no guarantee that any other body can succeed. We have in this study only the first step towards confrontation: a thorough examination by the church leadership of the cosmology of an African people. Hopefully, it will prove to be a seed-bed for the further and more difficult stage: suggestions and analysis on how Christianity stands with respect to fundamental cosmological ideas in Africa.

M. F. C. Bourdillon

INTRODUCTION

While I was busy writing this book, the last chapter in the short history of Rhodesia was written with blood, and missionaries had become apprehensive about their own future and that of their work. Some from their ranks had already become victims of the political and military struggle in the former colony, and the attitude of the regime in neighbouring Mozambique did naught for their comfort. Although this book is concerned with power, it is clear from its title that it does not set out to analyze the political situation of present-day Zimbabwe or explore the possibilities for its future development. It attempts to render a far more important service to the missions in that country, and beyond, which will retain its value even if future events should lead to a widespread interruption or total suspension of mission work there.

The first five chapters of the book undertake to expose power as the root concept of the traditional religion of the Shona in Zimbabwe.[1] The various spirits, upon whose actions the Shona believe that their life and well-being depend, are seen as symbolic representations, or conceptualizations, of those manifestations of 'power' which are looming large in their daily lives. It is my argument that Shona chiefs and spirit mediums, ancestral spirits and 'stranger spirits', witches and diviner-healers are subject to one basic notion of power whose chief characteristic is its ambiguity. In the rest of the book I show that this traditional conceptualization of spirits plays a central role also in the Shona 'Churches of the Spirit', which form an important branch among Independent Churches in Zimbabwe. It will be seen that they focus their attention particularly on those aspects of the Christian Gospel which meet the Shona people's traditional preoccupation with their spirits. This will become evident from my analysis of their doctrine, ritual and healing practices.

I hope this book will have a twofold effect. Firstly, to draw attention to those points in the Shona cosmology which will have to be confronted with the Gospel, if the latter really is to take root among the Shona people. Secondly, to encourage people elsewhere in Africa, particularly in Southern Africa, to undertake similar studies. Unless this kind of confrontation between traditional African cosmologies and the Christian Gospel is sought, there will never be a truly African Christian Theology. In the face of growing nationalism in Africa on a continental scale, and a concomitant move in many African countries to revalidate the traditional local religions, this confrontation is, thank God, in any case unavoidable.

Genuine African Christian Theology cannot be distilled from sentimental *négritude* literature, which abounds with generalizations about what is supposed to be typically African. It will only grow where, first of all, a hard-headed assessment is made of the connection between a particular African society's traditional institutions and its religious assumptions; and where, secondly, these assumptions are then confronted with the Gospel of Christ. This study is an attempt to present such an assessment with regard to the Shona. At the same time it shows, using the 'Churches of the Spirit' as an example, that where such confrontation does not take place, there is real danger of what might be termed a 'South Americanization' of the Church: superficially it would display a number of Christian semblances; intrinsically, however, it would remain tied to the traditional cosmology.

Generations of Whites and missionaries have looked down upon the bulk of religious notions held by the Shona and other African peoples, regarding them as naïve superstitions, although there have always existed notable exceptions. However, this book shows that the Shona belief in the existence of spirits is not simply superstition, in the sense in which this term is understood in present-day Western societies. Rather it represents a complete system of religious assumptions, a veritable *weltanschauung*, which is intended to answer man's most existential questions.

If the Church is ever to take deep roots among the Shona people and in their culture, her missionaries - whether they are Whites or Blacks – have to take the existence of this all-embracing traditional Shona religion seriously. It is true that in fulfilling her task, given to her by Christ, the Church has to absorb 'whatever good is found to be sown in the hearts and minds of men, or in the rites and cultures peculiar to various peoples'.[2] However, it is also her task to echo Christ's call to repentance through all generations. Genuine *metanoia* of a people and its culture will only come about as a result of the confrontation between that which 'people were told in the past' and Christ's 'but now I tell you'.[3]

This confrontation can only take place if on the part of the Church the existence of these African cosmologies – as fully-fledged religious systems – is taken seriously. Only then is it possible to appreciate to the full their importance in the life of the individual African. White missionaries, far from regarding a particular African religion as a systematic world-view, very often decry what little they have come to know about it as a curious collection of superstitious beliefs and practices. However, at the same time they must admit that most of their African converts have a foot in both camps: they have been baptized and go to Church, while they also cling tenaciously to essential points of the faith of their forebears. Because they have only a partial knowledge of the total indigenous religious system, White missionaries are also often surprised by the indifference with which their congregations greet some of their well-intentioned liturgical 'adaptations to the local culture'.[4] Being aware of the total scope of their traditional 'superstitions', and sensing that the Christian Gospel directs a similarly total claim at them, a merely superficial meeting between Christ's Gospel and the religion of their fathers leaves them dissatisfied. Instead of being surprised by their indifference, we should thank them for their sound intuition.

In statistics about the religious affiliation of the world population, Shona who have not joined one of the Christian Churches are classified as 'animists'. However, although to the Shona their belief in the activity of a variety of invisible powers in the world is central, 'animism' is an inadequate theoretical model for describing the dynamic character of their religion. In this religion man's basic preoccupation revolves around the question of how to acquire and to retain desirable power; and, where he feels threatened, how to keep undesirable power in check.

However, the Shona do not imagine power as a merely abstract entity. Even where power is thought to be immanent in inanimate matter it bears personal traits. Not only can it be used for achieving morally good ends, but it may also be misused for morally reproachable ones. This experience is reflected in the ambivalent attitude which the Shona display vis-à-vis the 'principalities and powers',[5] which form part of their cosmology. In a cosmos which is seen as subject to the law of power, there is no room for a relationship of trust between man and the 'powers' on which he believes himself to depend. Rather the resultant cosmology will aim at accommodating man with the 'powers', in an attempt to use them for his own purposes, or – where they threaten his existence – at neutralizing their influence over him and his environment.

In chapters six to eight I show that this traditional Shona view of the world as an enormous field of power has entered the doctrine and practice of the 'Churches of the Spirit' essentially unchanged. It is true that in their official doctrine these Churches ascribe to the Holy Spirit all those roles which formerly in the traditional Shona religion were held by different spirits or 'powers'. However, in their practice this has not led to a significant difference between their own rites and traditional Shona rituals. In addition, most members of these Churches see the alleged ultra-human powers of their Church founder vested in the latter's own person and not as an outflow of the Holy Spirit, as is their Church's official doctrine.

There can be no doubt that each encounter of the Christian Gospel with another culture necessarily leads to a *mélange*, as Daneel has expressed it very well in the title of his study *Old and New in Southern Shona Independent Churches*.[6] However, I believe that in the case of the Shona 'Churches of the Spirit', this *mélange* has put the Gospel at a disadvantage and consequently, has led to their 'South-Americanization'.[7] Putting the argument in Biblical language, one could say that in these Churches far too small a lump of yeast has been smothered by an excessive amount of flour.[8]

Over the past ten years, practical theology has become increasingly aware of the fact that the continued existence of certain traditional social or economic structures may effectively block the spread of the Kingdom of God among a particular people or in a whole geographic region. It is no coincidence that this awareness has developed most keenly in the ranks of the Church in South America. She accuses herself of having been linked far too long with the political heirs of the colonial past and of having only superficially Christianized the pre-colonial indigenous religions.

I hope that this study will make one important point very clear: mission work which neglects to address itself to the central tenets of a particular people's religion, and to judge these tenets in the light of the Christian Gospel, is bound to fail. However, before the traditional tenets can be confronted with the Gospel and judged by it, they have to be culled from their cultural wrappings. Only then is one able to grasp their theological relevance and develop a strategy for their integration into the life of the Local Church, or, where necessary, for their liquidation. As will be seen in this study, to cull the traditional religious tenets from their cultural wrappings means to show the intimate relationships which exist between a particular traditional religion and the social structures into which it is embedded. It is obvious that for this culling operation theologians cannot employ the methods of their own subject, but need to avail themselves of Social Anthropology as an auxiliary science. Therefore, in the presentation of my argument I adhere strictly to anthropological method and theory.

Dr M. F. C. Bourdillon, a distinguished anthropological fieldworker, whose studies on Shona religion are among the most perceptive available to date,[9] also views the Shona spirits as symbols of power. They are conceptual constructs through which man endeavours to account for the manifold 'powers' which he experiences as impinging on himself and the world around him. Abner Cohen, another social anthropologist, has pointed out that

> symbols are systematized together within the framework of dynamic ideologies, or world-views, in which the symbols of the political order are integrated with those dealing with the perennial problems of human existence: the meaning of life and death, illness and health, misery and happiness, fortune and misfortune, good and evil. These two symbolic complexes support one another within a unified symbolic system.[10]

Many years ago, Dr Godfrey Lienhardt of Oxford University made a now classic study of the religion of the Dinka, an African people living in the Sudan, and showed that their cosmology was just such a unified symbolic system as that to which Cohen refers.[11] The Dinka and Shona societies are separated by a distance of well over a thousand miles and there do not exist between them historical links that we know of. However, although the ecology of their respective habitats is rather different, Lienhardt's approach to describing Dinka religion can be equally well used for analysing Shona cosmology or that of any other people. Lienhardt's approach is succinctly expressed in the title of his book *Divinity and Experience.* I intend to show that the Shona people's traditional conceptualization of spirits derives from their belief in the existence of ultra-human powers and ultimate values, which for the individual and the group are intimately related to the experience of their total environment. There is, of course, a significant difference between the social anthropologist's conscious search for meaning behind the symbolic complexes which he discovers among the people who form the object of his study, and the people's own search for meaning through the unconscious employment of these symbolic complexes. The social anthropologist links the symbols with the people's history, ecology, and mode of survival, whereas the individual living in that particular society acquires his grasp of them largely by way of subconscious processes. Experience has proved to him and his forebears that these symbolic representations, taken in their entirety, provide a reasonably plausible explanation of the world – of its inconsistencies and paradoxes – although he may find himself unable to understand and explain them in detail.

Although Dinka and Shona differ from each other in a number of ways, there is a remarkable similarity between the symbolic statements which members of the two societies make with regard to 'powers'. Anthropologists who have written on the symbolic system of the Shona[12] do not seem to have found a word comparable to what in Dinka language designates the 'widest class of ultra-human agency', collectively called *jok*, Power.[13] Nevertheless, it is obvious from the result of his research that Lienhardt's findings with regard to Dinka 'Divinity' essentially reflects the assumptions which the Shona hold with regard to their own various 'powers': namely, that

> divinity is manifold as human experience is manifold, and of a manifold world. Divinity is one, as the self's manifold experience is united, and brought into relationship in the experiencing self.[14]

Without the 'powers', 'there would be for the Dinka no differentiation between experience of the self and of the world which acts upon it'.[15] By separating subject and object in experience, Dinka and Shona free themselves symbolically from what they must otherwise passively endure.

The Shona peoples and their neighbours

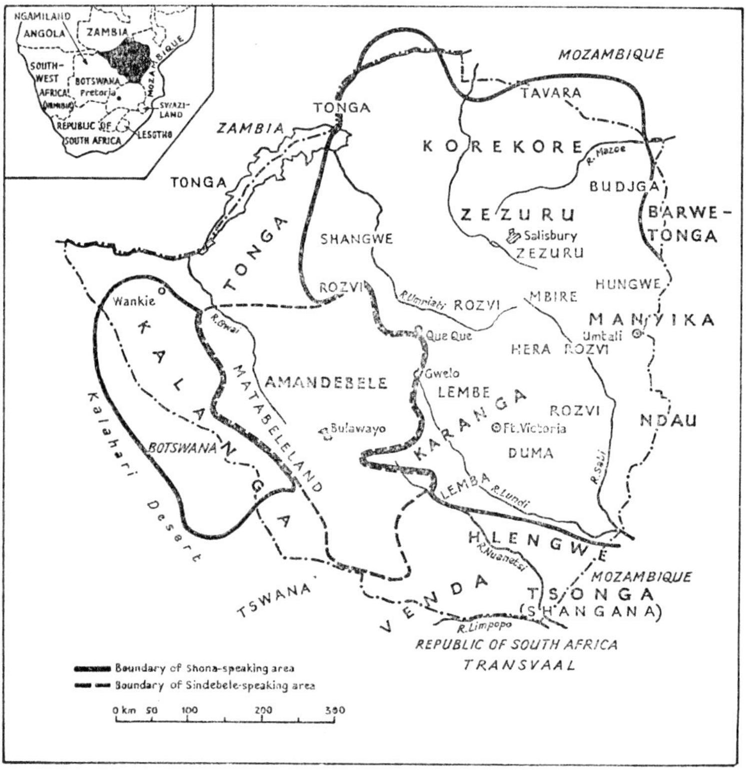

NOTES

1. Professor Kenelm Burridge in 1969 presented an interesting working definition of religion and religious activity, in an attempt to avoid the unnecessary wrangling with distinctions between religion and magic. For him, all religions are 'concerned with the truth about power'. Cf. Burridge (1969) p. 6f.

2. Cf. the Documents of the Second Vatican Council: Abbot (1967), p. 596.
3. Cf. Mt. 5,21.
4. Before 1971, Father Joseph Kumbirai, a Shona who has since left the Catholic priesthood, devised a Christian burial service which followed closely the traditional rites: cf. Daneel (1971), Appendix III. Kumbirai's liturgy was sent to the heads of all Catholic missions in Shonaland for experimentation and commentary. However, not only priests were divided in their opinion over the project: *ibid.*, pp. 270-7. Cf. also the readers' letters in *The Southern Cross*, of 19th September and 10th October 1976, under the heading: 'Why not christianize this Shona feast?' and 'This tradition can't be christianized.' The debate was about the traditional 'bringing-home' ritual which I describe on p. 76.
5. Cf. Eph. 6,12.
6. Daneel (1971 and 1974).
7. Interestingly, Dr. Peter Fry, to whom we owe a major study on spirit-mediums in Zimbabwe (cf. Fry 1976), now carries out research into pseudo-Christian Spiritist sects in Brazil.
8. Cf. Mt. 13,33.
9. Bourdillon (1972a and 1976).
10. Cohen (1969), p. 217.
11. Lienhardt (1961).
12. Apart from Bourdillon, cf. Aquina, Daneel, Fry, Garbett, Gelfand, Murphree, and Weinrich.
13. Lienhardt (1961), p. 31.
14. Lienhardt (1961), p. 156.
15. *Ibid.*, p. 170.

CHAPTER 1

THE SHONA PEOPLES

The peoples classified as Shona form the majority of the indigenous Bantu-speaking population of Zimbabwe.[1] The term 'Shona' is rarely used by the people themselves, who tend to refer to themselves rather by the name of the particular Shona-speaking group to which they belong i.e. the Karanga and Kalanga in the south and south-west, with the Ndau and the Manyika to the east of them; the Zezuru in the centre; and the Korekore and Tavara in the north. Each of these groups is itself composed of numerous subtribes.[2]

The Mutapa Empire

In comparison with what is available in the rest of sub-Saharan Africa, there exists an unusual body of archaeological evidence for the Shona region. From the tenth century onwards scattered references to these areas are to be found in Arabic manuscripts. Since the establishment of firm Portuguese bases at Sofala, and later at Sena and Tete, in the sixteenth century, a considerable amount of documentary evidence concerning events in this region is at our disposal.[3] Lastly, a wealth of oral tradition has been collected amongst the Shona people themselves.

From these sources it has been established that prior to the arrival of the Shona, there existed several successive Iron Age cultures in the area of present-day Shonaland, the earliest dating as far back as 300 A.D. Like the people who introduced the Iron Age into the Rhodesian plateau, the ancestors of the modern Shona-speakers are thought to have come from north of the Zambezi, possibly from the Congo basin. Their culture, which was more substantial than that of their predecessors, lasted from about A.D. 1075 to 1440. From this period date the first stone buildings at Great Zimbabwe. Most of the stone work of the 'Acropolis' to be seen at this magnificent site was completed during the period.[4]

From Great Zimbabwe, and a similar stronghold in the northern Transvaal, Mapungubwe, the new immigrants seem to have dominated the more numerous Iron Age people, without replacing the latter's culture completely. Alpers believes their dominance rested upon their superior organizational skills and he thinks it likely that they 'commanded new, powerful religious concepts and mystic abilities, for both Zimbabwe and Mapungubwe were ritual centres of outstanding importance'.[5] Their hegemony had also a firm economic basis, for they carried on a brisk trade in ivory, copper and gold with Arab traders plying the Sofala coast.

On the basis of oral tradition, it would appear that during the first half of the fourteenth century a new group of Bantu-speakers, the Rozvi, moved into Shonaland from the north, led by a mythical leader called NeMbire.[6] At the beginning of the fifteenth century emerged the first Rozvi Mambo,[7] Nyatsimba Mutota, bearing the praise name Mwene Mutapa. From 1420 to the middle of the fifteenth century, Mutota led a military campaign which resulted in the creation of a vast 'empire', bounded in the north by the Zambezi river; in the west by the Kalahari Desert; in the east by the Mozambique Channel down to the mouth of the Sabi river; and in the south extending across the Limpopo river into the northern Transvaal.

However, under the reign of Mutota and that of his son, Matope, internecine rivalries broke out among the junior members of the family whom the Mambo had appointed as rulers over his new vassals. Thus by the beginning of the sixteenth century, when the Portuguese began to establish themselves on the Sofala coast, Mutapa's 'empire' had already broken into a number of independent states, varying in size and importance. Of these the one ruled by Monomotapa,[8] with whom the Portuguese were in closest contact, was neither the largest nor the most important.

Recurrent meddling of the Portuguese in the internal affairs of the Mutapa kingdom led to its collapse by the end of the seventeenth century. By contrast, the Changamire Mambos of Rozvi stock, whose original centre of power was in legendary Guru-uswa, extended their dominance over the whole interior. Some think Guru-uswa roughly corresponds with present-day Matabeleland. In 1692 Changamire Dombo waged a campaign as far as the Zambezi Valley, causing the Portuguese to flee to the safety of Tete and Sena. However, disaster befell the Changamire in the early 1830s when their country was over-run, first by the Ngoni, and shortly afterwards by Mzilikazi's Amandebele. Both warrior-like tribes had fled from the rising Zulu empire of Shaka in the area of present-day Natal. The ruling house of Changamire was wiped out and the remaining Rozvi descendants were scattered all over the country. Mzilikazi installed himself in Bulawayo and his people took possession of what became known as Matabeleland. Both he and his son Lobengula, who succeeded him, claimed sovereignty over the entire area lying between the Limpopo and Zambezi rivers and bordered in the east by the Inyanga mountains. In order to press home their claim, they constantly sent their raiding parties against the many independent Shona chiefdoms into which the former Mutapa 'empire' had dissolved.

Colonization

From the middle of the nineteenth century, missionaries, traders and adventurers paid sporadic visits to the courts of Lobengula and various Shona chiefs. However, in 1888, the territory suddenly became involved in the European Powers' 'Scramble for Africa'.[9] To forestall a feared link-up between German-held East and South-West Africa; to prevent the land-locked Boer Republic in the Transvaal from gaining access to the sea via German South-West Africa; and to check possible Portuguese expansion from Portuguese East Africa, Britain proclaimed in that year 'an exclusive sphere of influence over all the Matabele domains'.[10] In 1889 Queen Victoria signed a Royal Charter establishing the British South Africa Company (B.S.A. Company), giving it authority to operate in Bechuanaland and beyond the Limpopo river, to the west of Portuguese territory, for the period of 25 years. Later, due to the outbreak of the First World War, the Royal Charter was extended by a further ten years.

That Cecil J. Rhodes's business and empire-building ambitions played a major role in this development is clear from the content of the so-called Rudd Concession, which had been signed by Lobengula the previous year and which formed the basis of the Royal Charter. It granted to the B.S.A. Company

> complete and exclusive charge over all metals and minerals contained in my [Lobengula's] kingdom, principalities and domin-

> ions, together with full power to do all things they [i.e. Rudd and his associates – representatives of Rhodes] may deem necessary to win and secure the same.[11]

In 1890, the Pioneer Column of White settlers was organized by Rhodes and set off from Kimberley to occupy Shonaland. Fortified posts were erected in places like Salisbury and Fort Victoria, and the Company's authority was extended to cover Manicaland also. In 1893, ostensibly to stop Amandebele raids into Shonaland, Matabeleland was invaded by Company forces. Following the defeat of Lobengula's warriors, Matabeleland was added to the other two territories to form what was to be known as Southern Rhodesia.

When the expected gold deposits proved less profitable than anticipated, the B.S.A. Company turned its attention increasingly to agriculture, assigning large tracts of land to settlers and mission bodies. To force Africans into providing cheap labour in mines and on farms, in 1896 the Legislative Council – composed of representatives of the Company and White settlers – imposed a personal tax of ten shillings per annum on every adult African. This and a number of other factors led to the rebellion of the Amandebele, which was joined by many of the Shona chiefs, some of whom were not subdued before the end of 1897. By the end of the rebellion, the Whites were in firm control of the highveld – undulating plateau country above 1 220 metres – which extends across Zimbabwe from the south-west to the north-east, where it is widest.[12] It is here and in the middleveld that most of the country's rich mineral deposits are found.

In 1902, with the establishment of the Native Affairs Department, the African chiefs' most important powers were transferred to European Native Commissioners, and their powers and duties declared to be no more than those of police constables.[13] Native Commissioners now allocated land to Black people, issued them with cattle permits and, from 1910, exercised criminal and civil jurisdiction over them. In the same year most of the boundaries of the 'Native Reserves' were also defined.

In 1922, B.S.A. Company rule came to an end. When Britain granted Responsible Government to the country, the new constitution made provision for racially segregated areas. In 1931 the Land Apportionment Act legalized the existing division of land between Blacks and Whites. This Act was followed by the Industrial Conciliation Act. Passed to protect White workers from competition by Blacks, it introduced in Rhodesia, for all practical purposes, South African style job reservation. During and after the Second World War, the country's economy experienced a rapid development which resulted in increased White immigration.

Between 1953 and 1963, Southern Rhodesia was amalgamated with Northern Rhodesia and Nyasaland, the three territories forming what was known as the Central African Federation. When it broke again into its component parts, Northern Rhodesia, as Zambia, and Nyasaland, as Malawi, gained their independence from Britain. In Rhodesia, the Federation period was marked by further economic expansion and to satisfy the increased demand for labour, the government now advocated partnership between the races. However, the lifting of certain colour restrictions benefited only a limited number of educated and wealthy Blacks, most of whom lived in the urban areas. The rural Black population was antagonized by the Land Husbandry Act of 1951, which had been passed to preserve soil in the 'Native Reserves' through conservation measures and destocking. Aiming at the creation of a stable rural and urban Black population – and thus preventing the continuation of labour migration – the Act also disqualified Blacks working in urban areas from cultivating land in their home villages.

Land Tenure

As in colonial Kenya and in South Africa, the unequal distribution of the land between Blacks and Whites in Rhodesia was sooner or later bound to become the cause of major grievances among the country's indigenous population. In 1920 slightly more than 210 000 hectares had been set aside for exclusive occupation by tribesmen, and this allocation was thought to satisfy their needs for an indefinite period. This impression was based on the estimate, that at the time, about one million Blacks lived within Rhodesia's borders.[14] In 1931, the Land Apportionment Act declared fifty per cent of the land as 'European Area', thirty per cent as 'Native Area', and the remainder as unassigned or forest land. The passing of the Land Husbandry Act twenty years later was an indication that overcrowding had already become a problem in many of the 'Native Reserves'.

Widespread dissatisfaction with the Act among Blacks eventually led to its abolition. It was replaced by the Land Tenure Act of 1969 which divided the land between Whites (1972:267 000) and Blacks (1972:5 490 000)[15] in the following manner (in hectares):[16]

Land Designation	White	Black	National
Forest area	737 682	177 663	
Parks & game reserves	1 793 353	255 146	2 676 062
General land	15 650 248		
Purchase area		1 484 998	
Specially designated	9 294	121 572	
Tribal Trust Land		16 151 521	
TOTAL	**18 190 577**	**18 190 900**	**2 676 062**

The Rhodesian government regarded the Tribal Trust Lands as the true home of all Black people. Urban Blacks who were unable to find employment in town were urged to return to their rural villages. Having to absorb these returnees put an additionally heavy burden on the Tribal Trust Lands.[17] Black Purchase Areas were blocks of land usually adjoining Tribal Trust Lands, in which proficient Black farmers were able to obtain freehold titles to farms. They were, however, considered as part of Rhodesia's cash economy, and to preserve the farms' economic viability, they could not be subdivided. For this reason, Black Purchase Areas brought little relief to the population pressure in the Tribal Trust Lands.

Weinrich found that the annual income of a farm owner in a Black Purchase Area was about four times as high as that of a peasant cultivator in a Tribal Trust Land. However, she notes that the annual incomes of £193 and £47 respectively are both small when the large number of dependants is taken into account. Peasant families, she says, often find it impossible to live on the income which they derive from agriculture.[18] For this reason, in the Tribal Trust Lands which she surveyed, most men aged between fifteen and fifty were away from their homes as migrant labourers.[19] If one takes into account that the growth rate of the Black population in Zimbabwe Rhodesia has shown an increase from 2,8 per cent in 1948 to 3,6 per cent in 1969 – resulting from a birth rate of 52 per 1 000[20] – one has to agree with Kay who wrote that

> the real problem areas of Rhodesia [are] the overpopulated, overstocked Tribal Trust Lands which have had to absorb 75 per cent of the total increase and in doing so have added something in excess of forty per cent to their 1962 population . . . In these areas lie Rhodesia's poverty and hunger, largely hidden from the eyes of most townsfolk.[21]

A natural concomitant of this situation is that an increasing number of Blacks gravitate to the towns. Between 1962 and 1969, this led to a growth rate of twenty-three per cent in the urban Black population, although the increase varied from town to town, ranging from as little as ten per cent in Umtali, to as much as 108 per cent in Que Que.[22]

Towards Zimbabwe

Native agitators became increasingly active in the Black reserves in the 1950's as a result of the increased contact between Southern Rhodesian Blacks and Black nationalists in the other two territories belonging to the Central African Federation, and because of the widespread dissatisfaction with the Land Husbandry Act. In response to this new development, the Southern Rhodesian government changed its policy towards the Black chiefs. Whereas until then Native Commissioners had ignored them completely, from the early 1950s onwards the chiefs' position was revalidated; their salaries increased, and they were organized into provincial assemblies allowing them to bring their views to the attention of government officials.[23]

The new policy soon began to show the hoped-for results. In 1959, the chiefs, through their provincial assemblies, openly sided with the government, asking for the suppression of meetings organized by the African National Congress Party in their areas. This party had been founded in 1957 and the discontented rural population looked increasingly to its leaders, rather than to the chiefs, for guidance.

Tribal areas and places mentioned in this book

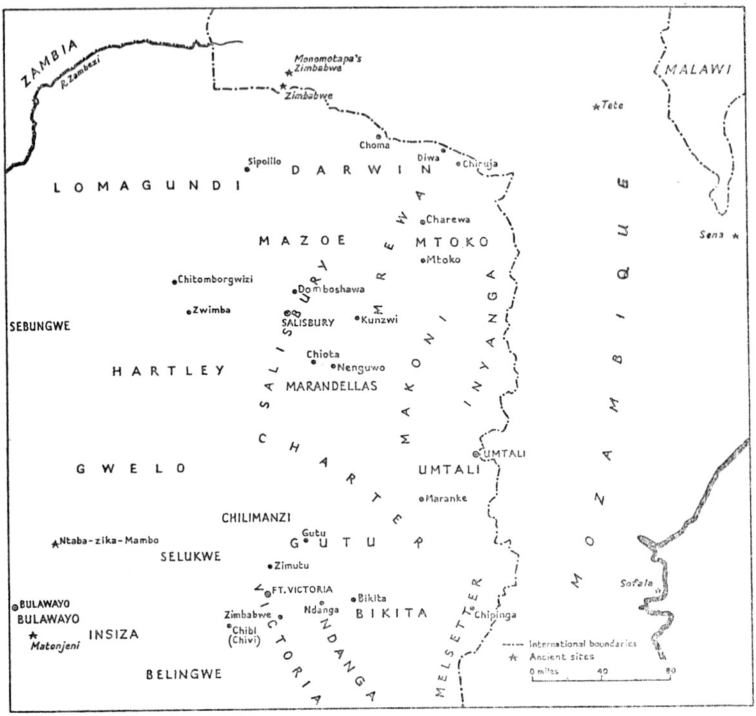

Having been banned several times, and each time re-emerging under a new name, the Black nationalist movement eventually split and throughout 1963 adherents of the two rival factions were locked in a bitter feud which led to the death of many people. Both factions, the Zimbabwe African People's Union (ZAPU), led by Joshua Nkomo, and the Zimbabwe African National Union (ZANU), under the leadership of Ndabaningi Sithole, were banned in the same year and their leadership interned.[24] An uneasy peace returned to the rural areas thereafter and the chiefs remained the only African leaders recognized by the Rhodesian government. In 1962, in the same year in which the White-led Rhodesian Front Party came into power, the authorities further enhanced the status of chiefs through the formation of a National Council of Chiefs.

Following Rhodesia's Unilateral Declaration of Independence in 1965, the power of chiefs was consistently increased, until it could be said of them that they had now 'completely changed their image: from leaders of rebellions at the end of the nineteenth century, to parliamentarians in a racially composite society'.[25] However, the majority of the African population in both town and country stood behind the nationalist leaders and supported the guerrilla campaign which these waged against Rhodesia, with the aim to create a new 'Zimbabwe'.

NOTES

1. According to the latest census, there were 3 433 960 Shona-speakers in Rhodesia in 1969. The next biggest language group were the Amandebele, numbering 706 750 people. Besides these two major groups, there are a number of smaller African tribes in Rhodesia, such as the Venda, Tsonga (Shangana), and the Tonga, who in 1969 together accounted for roughly a further million people.
2. Akers (1973), p. 362f. See also Map 1.
3. In this subsection I depend mainly on Alpers (1968).
4. For the most up-to-date treatment of the Zimbabwe 'mystery', cf. Garlake (1973).
5. *Op.cit.*, p. 6.
6. Abraham, who collected most of these traditions, takes them to be literally true: cf. Abraham (1959). Alpers (1968, p. 8) thinks it possible however, that the Rozvi were one of the families of chiefs already in the country, who rose to prominence when the ruling dynasty's powers began to decline.
7. 'Great chief' or 'King'.
8. The Portuguese rendered the Mambo's praise name, Mwene Mutapa, as Monomotapa.
9. Cf. the title of Nutting (1970).
10. Davies (1975). For the following I depend mainly on this source and on Weinrich (1971).
11. Davies (1975), p. XVII.
12. The other distinctive geographic divisions of the country are; the middleveld, between 915 and 1 220 metres which flanks the highveld to the north-west and south-east, and the lowveld, largely below 600 metres. The latter forms the Sabi-Lundi and Limpopo basins in the south and south-east and the Zambezi basin in the north-west. A fourth division is formed by the mountainous country, largely above 1 800 metres, which runs from north to south along the eastern border.
13. Cf. Weinrich (1971), p. 10.
14. Hughes (1971), p. 61.

15. Cf. Davies (1975).
16. Akers (1973), p. 205. 'General land' was also referred to as 'National land'. At the beginning of 1979 the Land Tenure Act was repealed. However, since at the time most of the country was ruled by marshal law, making a change of the *status quo* impossible, the description of the effects of the Land Tenure Act still holds true.
17. The population density is, of course, not equal in all 166 portions of Tribal Trust Land. Weinrich's nine sample communities had a density of between 120 and 237 persons per square kilometre.
18. Weinrich (1971), p. 36.
19. *Ibid.*, p. 37. In the Black Purchase Areas which were included in her sample, the same applied for most men between twenty and thirty-five years of age. According to her, 43 per cent of all men from Tribal Trust Lands and 33 per cent from the Black Purchase Areas work away from home as migrant labourers.
20. Hawkins (1974), p. 8. The UN-calculated average growth rate for Africa is 2,5 per cent. In 1969, fifteen per cent of the Black population were under sixteen years of age.
21. Kay (1972), pp. 29 & 31. Between 1962 and 1969 the population in Tribal Trust Lands had increased by 46 per cent, numbering 2 911 000 people, and in Black Purchase Areas by 25 per cent, to 132 800 people. Cf. *ibid.*, p. 28.
22. *Ibid.*, p. 29. In 1969, Salisbury had the largest Black population, numbering 340 000, followed by Gwelo and Umtali with 36 000 each, Que Que (29 000) and Fort Victoria (8 500). Cf. Akers (1973), p. 412f. I leave out the figures for towns in the Amandebele territory.
23. I ignore the history of the 'Native Boards' which were founded on government orders in 1931 to counteract the first – rather moderate, by modern standards – Black political associations which had then made their appearance among urban Blacks. Their further development is described by Holleman (1968), Weinrich (1971), and Howman (1959).
24. Loney (1975), pp. 99-108.
25. Weinrich (1971), p. 22.

CHAPTER 2

'OWNERS OF THE LAND': CHIEFS AND TERRITORIAL SPIRITS

For the Shona, as for people elsewhere in sub-Saharan Africa, land is not merely an economic asset, but has a value which is intimately linked with the tribe, its chief, and the spirits of their ancestors. The chief is often referred to as the *muridzi* (or *mwene*) *we nyika*, the 'owner of the land'. The meaning of *nyika*, the Shona term commonly used for 'land', comprises also the chiefdom and the people, for '*nyika* means people occupying the land, and their rights thereto', and there is a Shona saying that 'the chief is the people, the people are the chief'.[1] There are also numerous myths in which these intimate relationships find their expression. In one of them, the chief of Matonjeni, a place in the Matopo hills near Bulawayo, when he was distributing his estate among his sons, is said to have given a purse of soil to one of them, and a purse filled with grass to the other, whereas the rest of the brothers received his cattle. Naturally, the two who received the strange gifts were laughed at by their cattle-rich brothers. However, their nephews[2] reassured them that their father had in fact handed to them, in a symbolic manner, the right to succeeding him in the chieftainship.[3]

In a number of Shona chiefdoms even today, the installation ritual of chiefs contains a ceremony which seems to be bound up with the *leitmotiv* of this myth. In the Chitomborgwizi Black Purchase Land Area, the chief-elect has to undergo the following ordeal: Earth which has been dug out from underneath a bush is given to him and he has to hold it in his clenched left fist. Into his right hand is put some maize-meal, mixed with small seed grains. Holding both hands tightly closed, the chief then enters a deep pool, together with his two masters of ceremony, who immerse him into the water for some moments, before returning with him to the dry land. In this way, it is impressed on the future chief that he is about to assume responsibility for his territory and all that lives and grows on it.[4] Bourdillon reports that among the Korekore the medium of the territorial spirit (*mhondoro*) gives to the chief-elect a handful cf soil which has been mixed with body fluids of the late chief or with soil from his grave, and with other 'medicines', in order to 'make him strong'.[5]

The chief's special relationship to the land also becomes apparent from the widespread taboo which forbids commoners to eat the flesh of an antbear – which burrows in the ground – while it is a delicacy for the chief. Likewise, the flesh of that side of an elephant on which it lies when it drops dead is reserved for consumption by the chief's household.[6]

Powerful founder-ancestors

The chief's 'ownership' over the land and its people derives from his supposed connection with the mythological founder-ancestors of his chiefdom. It is they who are believed to have chosen him, and bestowed on him their own prodigious powers. In their own lifetime they are thought to have possessed powerful 'medicines' which enabled them to overcome the original owners of the land, and, subsequently, to defeat any aggressors who attempted to oust them from their newly-acquired possession. Frequently, however, the myths reflect the fact that in reality the occupation of a new territory may have taken place in a rather more peaceful manner, the original occupants moving away to other lands, or continuing to live among the invaders. Thus the original 'owner' of Mtoko district, Makate, was overcome by Nehoreka, after the latter had persuaded him to accept his sister as a wife. Through her, Nehoreka got hold of Makate's powerful charms which had enabled him to repulse any previous attackers. Makate was, however, not defeated, because after discovering the theft he led his people into a rock. The myth asserts that when Nehoreka's people had taken control of Mtoko, Makate's tribe continued to live among them, invisibly.[7]

A mythological theme which recurs in many different contexts is that the tribe's founder-ancestor acquired his extraordinary powers by committing incest with his sister. One author claims that among the Korekore, it was thought to be customary 'until fairly recently' for successors to the great chieftainship 'to fortify their installation', i.e. to ensure their own fertility and that of their chiefdom, by committing ritual incest with one of their sisters.[8]

In the same connection, one encounters again the theme of the founder-ancestor's disappearance into a rock. Nehanda – venerated by the Korekore as their female territorial spirit – was so disgusted after she discovered she had slept with her half-brother Matope that she retired to her country and disappeared into a small hill, 'where the outline of the apron which was used during the ceremony can be seen impressed into the rocks'.[9]

One of his informants pointed out to Bourdillon that Matope 'did what others were afraid to do', and he comments: 'because he did what others were afraid to do socially, it is presumed that he could [even] do, what others could not do, physically'.[10] There is also a popular belief among the Shona that some people obtain good crops by committing incest,[11] and a Shona writer states that 'for medical purposes a brother and a sister may mate'. According to him, committing incest is the only cure for 'a victim of a fluke which has penetrated the skin and entered the body'.[12]

Even present-day chiefs are widely thought to possess powerful charms, a belief which at times brings on them the suspicion that they make use of witchcraft to achieve their ends. At the conceptual level at least, chieftainship for the Shona is still associated with extraordinary power. Apart from those factors already mentioned which are believed to account for the chief's powers, there are of course, other more 'pragmatic' ones affecting his status among his people. If he had a weak personality, a chief would not be highly respected, nor would he be if he were a poor man. There have been cases where men have been turned out of the chiefly office because they were not able to meet the expenses of ritual, hospitality and generosity which are demanded of its holder.[13]

For the Karanga, their chiefs are guardians of two fundamental values sustaining the chiefdom, i.e. *rupenyu* (life) and *simba* (strength, power, violence).[14]

> [The chief] possessed greater *simba* . . . than any of his subjects because he acquired it during the succession ritual . . . He was the only source of security which the people knew. Land was and still is . . . *rupenyu* for them, and the chief is the . . . owner of the land and therefore the guardian of life. His ability to retain a large following depended to a large extent on his control over sufficient land for his people . . . From these basic values, *rupenyu* and *simba*, and their material expressions, land and cattle, people derive their criteria for judging whether a chief is good or bad.[15]

It is in the light of this statement that Bullock's observation, made fifty years ago, can be seen in its true perspective. He tells of a man who, when he was rebuked by a chief for not saluting him, retorted: 'Where is the land?' Clearly, what the offender implied was that the chief had lost his title to rulership when his effective control over the land had been taken away from him by the European settlers.[16]

Territorial Spirits and spirit cults

The real 'owners of the land' are, of course, the spirits of the deceased tribal rulers and particularly those of the mythical founder-ancestors of the chiefdom, I have already mentioned that during the installation of a Korekore chief it is the medium of the territorial spirit who hands him the lump of soil with which he symbolically assumes responsibility over the chiefdom. According to Garbett, all land occupied by the Valley Korekore, whether or not it is inhabited, is divided into what he terms 'spirit provinces'. These provinces consist of areas varying in size and each of them is associated with, and often named after, one of the first Karanga invaders or, occasionally, an original inhabitant. The spirits of these men are believed to protect the fertility of the land and to control the rainfall. The territorial spirits of the different provinces are brought into relationship with each other by means of long genealogies. Some of them have 'come out' in mediums[17] who are identified with them and hierarchically ranked in accordance with the genealogical seniority of the territorial spirit whom they represent. At the apex of the hierarchy there are several senior mediums, each of whom represents a particular founder-ancestor. They each have authority over a number of lesser mediums whose appointment they must confirm.[18] There are, however, several spirit cults whose influence goes much further than that of the most senior territorial spirit of any particular 'spirit province'. One such cult is that of Dzivaguru and Karuva in the north-east of Korekore country. Although people normally honour only the territorial spirit of the chiefdoms in which they live, during prolonged droughts delegations from chiefdoms over a hundred kilometres away come to the sacred grove of these two famous spirits to ask for rain for their home areas. In central Shona country, the most powerful spirit mediums are said to be possessed by the spirits of people who lived in the country so long ago that nothing about them is remembered. Prayers are addressed to the most important of these spirits, Chaminuka and Nehanda, in rain rituals in widely dispersed localities.[19]

The best known of all the territorial cults is that of Mwari, which is dominant among the Kalanga and Karanga in the south of Shona country. However, its influence extends as far as Manyika and Ndau territory in the east, and into Vendaland in the south-east, over 400 kilometres from the cult's centre in and south of the Matopo Hills.[20] The officials at the main shrine in Matonjeni keep themselves informed about the state of affairs in a large number of chiefdoms by maintaining a network of over forty resident messengers who pay regular visits to Matonjeni. On their visits the messengers bring with them gifts for Mwari, consisting of token payments made by the heads of all the homesteads throughout their district in recognition of Mwari's being the 'owner of the land'. The shrine consists of a cave from which the 'Voice of Mwari' speaks its oracles to the messenger who must sit during the seance with his back to the cave. The 'Voice' and its intermediaries, i.e. the highest officials in the colony of priests and priestesses looking after the shrine, use the seances to elicit from the messengers information concerning their respective territories, and to send messages through them to the chiefs and people.

Occasionally, delegations come to Matonjeni to consult the 'Voice' on matters of public importance, such as the appointment of a new chief, the verification of the 'coming out' of a new spirit medium, drought, or some other communal disaster. On these occasions the oracle frequently exhorts the people to remember the traditions of their ancestors and warns them against embracing the new religion brought to them particularly by the Shona Independent Churches.[21]

Until recently, a persistent tradition dating back to Carl Mauch's speculations about the origin of the Zimbabwe ruins[22] claimed that the Mwari cult had held a central position in the cosmology of all Shona peoples for a long time. It is still echoed in Daneel's *Old and New in Southern Shona Independent Churches:*

> For centuries the Shona have believed in Mwari as the final authority above and behind their ancestors, a High-God who seemed perhaps less directly involved in their individual lives than their ancestors were, but one who could be consulted on matters of national importance.[23]

It seems, however, that the recognition of Mwari as Shona High God is of relatively recent origin. According to Ranger, new archaeological and oral evidence indicates that the Mwari cult is native to the Matopo Hills and the adjacent areas and that it has been embraced by successive civilizations living there. It seems to have been spread from the Matopos to Karangaland by the Changamire Mambos of Rozvi stock, whose dynasty lasted from the middle of the seventeenth century until the Amandebele invasion during the early nineteenth century.

Like their Rozvi predecessors, the new rulers left the cult organization untouched, incorporating Mwari into their own cosmological system under the name of Molimo.[24] Bourdillon points out that people seek the help of Mwari's oracle at Matonjeni in those situations of need in which Shona elsewhere petition their regional spirit mediums,[25] and Daneel admits that 'to this day the messengers who move between their home districts and Matonjeni still regard the petition for rain as their main assignment'.[26] As they approach the oracle's cave they call out Mwari's most popular praise name, *Dzivaguru* (Great Pool), which is obviously associated with the idea of rain.

Present-day Shona are said to believe in a remote high god, whose names range from '*Dedza*' (Lord in the sky) to '*Mutangakugara*' (The one who existed at the beginning). However, people often merge and confuse the widely known territorial spirits such as Dzivaguru and Chaminuka with the concept of the high god. Bourdillon makes the discerning statement that 'Mwari is *now* the most common name for the high god'.[27] He is on safe historical ground when he states that its wide acceptance today among the Shona is mainly a result of the adoption of Mwari by the Christian missionaries a hundred years ago as the name for the Biblical God.[28]

Comrades-in-arms

Since both chiefs and spirit mediums derive their authority from the territorial spirit, it is not surprising that their powers overlap considerably, and support each other, although Bourdillon emphasizes that 'the religious realm is primarily concerned with power over nature for the ordering of such natural events as rain, wind and the spread of pests, while the political realm is primarily concerned with power over people for the control of social events.' However, he also stresses that 'just as the social order is affected by the natural order, so political power is affected by religious power'.[29] Thus, one has to add the spirit mediums to the list of factors already mentioned, which have a qualifying effect on the power and status of a chief. A chief in whose area a powerful spirit medium resides has to be careful to avoid incurring negative sanctions of the territorial spirit for disobeying his orders. Chief Sileya, for instance, says that such disobedience would cause the whole country to dry up.[30]

Ideally, the chief and the spirit medium should be comrades-in-arms. Garbett says that a person wishing to establish a new hamlet must first approach the chief of the area and give him a small gift. The chief then points out a site which the newcomer may occupy. However, before developing the site, the newly-arrived has to go to the spirit medium representing the territorial spirit in whose province he has settled and make an offering of black and white cloth. If the territorial spirit is favourably disposed towards his wish to settle in the area, the medium will give him a 'medicated' pole which he has to build into the structure of his hut.[31]

Bourdillon observes that the annual rite of 'tasting the first fruits' 'clearly expresses the unity between the chief and the [territorial spirit] mediums, emphasizing the fact that all power is believed to come from the same source, i.e. the founding-ancestors of the chiefdom'.[32] During this ceremony, chief and medium partake together of specially 'medicated' porridge and beer, made out of grain from the previous and the present harvests. Similarly, both representatives of the founder-ancestors' power are expected to co-operate in times of calamity affecting the whole tribe and on occasions when the entire chiefdom gathers in thanksgiving.[33] The existence of this type of co-operation between Shona chiefs and spirit mediums was already observed nearly four hundred years ago by the Portuguese writer Dos Santos, in the Karanga territory which at the time was ruled by a chief Kiteve.[34]

Dos Santos first describes the ritual which was held each year in September, at the appearance of the new moon, when Kiteve and the people would ascend a high mountain near his residence, the Zimbabwe. Apparently it had a similar character to the rites performed elsewhere in Southern Africa, with a view to 'renewing' and 'strengthening' the soil.[35] 'After the festivities', Dos Santos writes,

> They had a great weeping for the dead, which lasted two or three days, "until the devil enters into some Kafir in the crowd, who declares that the soul of the dead king in him wishes to speak to his son or successor". On these occasions the chief would lie on the ground like a man possessed, speaking the dialects of many tribes. Then he would begin to splutter, and to talk with the voice and the manner of the late chief. The Kiteve and his induna at first lay prostrate before the operating Kafir; and then the Kiteve, left alone with him, would consult him about his future prospects. The supposed spirit of evil would fill the chief's mind with lies, but that would not prevent a similar consultation every year.[36]

If the chief's status and power are affected by the presence of an influential spirit medium in his chiefdom, the obverse is, of course, also true. This is well documented by the reaction of the people of the Lomagundi area to the arrival of a pioneer Wesleyan preacher in 1892, called Shimmin. They assured him and his company that 'they were all eager to learn the white man's faith, but that they could not act in the matter until the consent of their great prophetess had been obtained'. He was told that she could make rain, govern the harvest, avert sickness or bring punishment, and that 'to disobey her voice means destruction for the present and the future'. The chief, therefore, sent a messenger to the spirit medium to request her permission to have the new teaching established amongst them. After a few days she replied that 'so far as she was concerned she had no objections, but that in a matter of this sort she felt she should consult with Lobengula', the paramount chief of the Amandebele. The latter regarded all Shona as his subjects, though his residence, Bulawayo, was a month's journey from the Lomagundi chiefdom.[37] There is also the case of a Zezuru chief who ordained that his people may not consult the spirit medium residing in his chiefdom without having approached him first.[38]

Rivalry

The overlapping of the powers which are attributed to the chiefs and spirit mediums, and the fact that these powers are believed to flow from one and the same source, also provide scope for tension and open conflict between the two holders of power. Understandably, such tension is less evident where the mythical senior territorial spirit is both the 'owner of the land' and the founding-ancestor of the chiefly dynasty, which is the case in most of the Shona chiefdoms. There are, however, exceptions. In Choma, in north-eastern Zimbabwe, for instance, the chiefly dynasty is distinct from the agnatic group of the senior spirit medium. The two lineages find themselves, therefore, in an ambivalent position, the chiefly descendants of Chigango being ritually subject to Karuva, the senior territorial spirit in the area, while the descendants of Karuva are subject to the chiefly powers of Chigango's lineage. According to Bourdillon, each of the two lineages establishes its seniority over the other by claiming to have descended from the original 'owner of the land'. They support these claims through the manipulation of a common pool of tribal myth, with the result that the essentials of the two versions as told by the two vying parties are diametrically opposed. In the chief's account, Chigango was the first occupant of the chiefdom and the 'owner of the land', whereas in the second version Karuva was the first to live in the chiefdom and Chigango took the chiefdom only at his concession.[39] This kind of rivalry is particularly rife within that area of southern Shonaland in which the Mwari cult is widespread and represented in each chiefdom by a resident messenger of the cult. Thus in 1892, the Berlin missionary, Beuster, witnessed a disagreement between chief Chivi and a Mwari cult officer over the payment of tribute to the central shrine. It is reported that Mwari's representative 'won his point by threatening to organize an Ndebele raid, but for the time being the popular feeling was for the missionary and against the Mwari cult'.[40]

The political power of the spirit mediums is most clearly observable in the fact that it is they who in effect elect the ruling chiefs, although their choice is, of course, believed to be that of the territorial spirits, who use them as their mediums. Since the Shona rules of collateral succession are such that chiefs are invariably fairly old men when they are installed, and since spirit mediums, on the other hand, receive their calling at a rather young age, it follows that each of the latter, in the course of his life-time, is likely to be involved in the election of several successive chiefs. When a man who has established a chiefdom dies, each of his sons, provided he is not mentally or physically incapacitated, ought to succeed in turn. On the death of the last male of the filial generation, the eldest grandson ought to succeed, followed by the oldest surviving son of the founder's second son, etc. After all eligible men of this generation have succeeded, the title ought to drop a generation and rotate in the same manner among the descent groups established by each of the founder's sons in order of seniority.[41]

As Latham points out, with this system of succession, problems arise already within a generation or two, given the fact that classificatory kinship terminology is used and families are polygynous. Confusion over generation levels is an almost immediate effect and, in the absence of birth registers, considerable argument may arise over the relative age of collaterals.[42]

In each Shona chiefdom there exist therefore, several houses which are contending with each other for the chieftainship. Often the contenders are influential headmen, or at least kraalheads, and serving as councillors in the ruling chief's court. Although the territorial spirits are believed to kill usurpers of the chieftainship, or to cause them to go mad, and the same punishment is said to be meted out to imposters who claim to be the medium of one of them, it is quite clear that succession disputes are bound to be complicated and wide open to machinations behind the scenes.[43]

Chiefs' quest for power

It is in the light of the chiefs' dependence on the goodwill and co-operation of the spirit mediums, and their potentially weak position vis-à-vis the bulk of their subjects, that one must understand their quest for means to increase their power. In this connection, some instances from the early contacts between them and the advancing Whites during the latter half of the 19th century are very illuminating.[44]

In his account of the first Protestant missionaries who came into contact with Southern Shona chiefs, Beach writes that 'they did not have sufficient success to indicate that a desire for conversion was the real reason for this reaction'. Rather, the true reasons for the welcome they received were their coveted trade goods and their admired marksmanship. One of them wrote that 'the captains were very greedy; if they heard that we were moving to other captains . . . they would think that the others would get more benefit and become more important'.[45] Less than half a century later, in 1888, when the British South Africa Company had begun to cast its shade over the country, the first Anglican bishop of Mashonaland found more than 30 chiefs willing to receive Christian teachers among their people. At the same time, their overlord Lobengula, realizing his weak position as a ruler, due to the fact that he was not his father's rightful heir, entered into an alliance with White settlers in the hope of gaining firm support.[46]

Of chief Nenguwo, at whose village the first Methodist Training Institution for Shona teachers was established, it is expressly stated that he was the only chief in the area who was prepared to accept the missionaries. The internal cohesion of his small chiefdom had been weakened by a series of succession disputes, and its position between two powerful chieftaincies with a long history of feuding, posed a continual external threat.[47] It is also interesting to note how the Shona attempted to account for their first experiences with the extraordinary power of the advancing Whites by assimilating them into their mythological lore. Thus it was said that the earlier mentioned 'owner' of Mtoko district, Makate, had been a White person in disguise, who only pretended to disappear into the rock, whereas in fact he went back to Europe for more charms and friends. This is why Whites came here soon afterwards. On their second coming to Rhodesia, Makate and his people no longer had any reason for disguising themselves.[48]

The Shona chiefs' earlier eagerness to avail themselves of the strangers' imposing power for the achievement of their own ends was soon followed by the dispiriting discovery that they were dealing with 'reckless land-grabbers and spoilers'.[49] As late as 1960 a despairing lament for the ruin created by the white man and the powerlessness of the defeated, speaks of the 'spoiling of the land':

> Ay, this world fails one,
> Ay, the unburnt pot has entered here,
> Ay, the unburnt pot has spoilt the world.
> He has cut the world in the middle . . .
> The unburnt pot has entered the world and killed it . . .
> Yelele, the unburnt pot just handles the world,
> twisting it.[50]

In traditional Shona thought, the concept of 'spoiling', or 'polluting', the land – exciting the wrath of the territorial spirits and causing them to withhold the rain and other blessings – occurs in connection with the breaking of a variety of social and ritual taboos, such as incest and homicide; the burial of deformed or stillborn babies, or of women who died in labour and were buried with the foetus still in the womb. The concept occurs in general with any unusual event, as, for instance, twin births, or babies who cut their upper teeth first.[51] The Whites' brushing aside of the traditional modes of ownership, the institution of their own administration, and their suppression of what they regarded as primitive superstition, has completely 'spoilt' the land. It is not surprising, therefore, that in the Shona risings of 1896 the spirit mediums took a leading role, as will be seen later. Ranger observes that the risings were characterized by an obvious ambiguity towards White ideas and technology: 'there is repudiation but also desire; a rejection of White mastery, but a longing for African control of modern sources of wealth and power in an African environment.'[52]

Social anthropologists and historians have brought together an impressive documentation on the subject of the 'spoiling' of the chief's and the spirit medium's power in present-day Zimbabwe. Since this phenomenon has a direct bearing on the emergence of the Independent Churches and some of their characteristics, I have to treat it in some detail, and will do so in the remainder of this chapter.

The 'spoilt' land

Garbett believes that

> probably the most far-reaching changes in the political processes operating within the chiefdoms were brought about by the creation of the Reserves, and by the Native Commissioner becoming responsible for appointing a chief's successor.[53]

The spirit mediums are no longer the sole selectors and installers of chiefs. Even though the Commissioners, as a rule, leave the process of electing a new chief in the hands of the tribal institutions, they are empowered to reject the spirit medium's choice and have done so fairly often. Sometimes this has led to two men in the same chiefdom claiming to be the rightful chief. As has been seen, it has also resulted in succession disputes becoming even more complicated. The severity of succession disputes has intensified because disgruntled contenders for the chieftainship can no longer secede and form new chiefdoms, due to the permanent fixing of boundaries against white farmland and within the Tribal Trust Lands.

Chiefs are nowadays salaried by and responsible to the central government in Salisbury. As Ndabaningi Sithole, leader of the Zimbabwe African National Union, put it, they 'no longer represent the will of the people, but that of a foreign power . . . European powers, while preserving the shell of African chieftainship, have emptied . . . [it] of its real content'.[54]

The question as to how much real power the Shona chiefs still have over their people, has assumed greater political importance after the Rhodesian government's unilateral declaration of independence from Britain. Before the declaration was made, the government called a meeting to which it invited all chiefs, headmen, and kraalheads. It is commonly referred to as the Dombashawa Indaba.[55] After the meeting it was claimed that the overwhelming majority of chiefs backed the intended declaration of independence. However, a White leader of long-standing, Sir Godfrey Huggins, at the time rejected the Rhodesian Front's claim that the chiefs were the real leaders of the Shona as a 'swindle'. 'Many of the Shona chiefs', he said, 'are rather dodderers. Their real powers and influence were destroyed at the time of the rebellion in 1896'.[56] This view is also reflected in the Report of the Pearce Commission drawn up in 1972, which came to the 'inescapable conclusion' that

> the African people, whether rural or urban, do not recognise the chief's authority in matters outside tribal custom, and inasmuch as most chiefs rejected the Proposals [for a settlement between the Smith Regime and the British Government], this could be a reflection of their real role, as interpreters rather than arbiters of the public will.[57]

Nevertheless, there do, in fact, still exist some chiefdoms where chiefs have managed, due to a variety of factors, to uphold their traditional authority. Weinrich presents the case histories of two chiefs, one of whom is a confirmed conservative, but who is also wholly co-operative with the White administration, and whose authority in the eyes of his people is enhanced by the belief that he is host to two territorial spirits. The other has managed to toe the line of the government, flirted for a while with the Catholic Church, endeared himself to the nationalists, and restored his chiefdom's ancient association with the Mwari cult in the Matopos.[58]

It is obvious that if the country is going to be turned into a modern nation-state based on the Rhodesian African nationalists' ideals, there will be less and less room left for chiefs and spirit mediums to hold sway over the people who live in their territory. While the chiefs, at best, could hope to function in that new situation as civil servants, as is now the case, the spirit mediums would seem to become totally dispensable. Such a turn of events has, however, not yet taken place. It is therefore necessary to take a closer look at the reasons why the spirit mediums' powers have, on the whole, better survived the onslaught of modern times on the traditional system than have those of the chiefs.

I have already noted the decisive influence the spirit mediums have on the election of new chiefs and it will be seen shortly that they are also usually considered as the sole authorities on the history of their chiefdom, although this claim does not always go unchallenged. By far the most important reason for their continued influence has been well stated by Garbett:

> Given the Korekore belief in the powers of the spirit guardians [= territorial spirits], it may be that the persistence of the organisation based upon the spirit mediums is due not only to the isolation of the region, which has served to preserve the belief, but also to the vagaries of the climate and the poor physical environment; *Korekore are faced continually by the threat of drought and famine*.[59]

Bourdillon writes that in the eastern Korekore chiefdoms where he did his research, tension mounts among the people around the beginning of November, when the expected rains dominate conversation. The unease remains until the crops are well established and grows into anxiety should the rains, which tend to follow an erratic pattern, falter before this.[60]

The Amandebele (Matabele) Rising of 1896 was provoked by a phenomenal combination of physical plagues falling upon the country, which were interpreted by the Karanga priests who served the Molimo[61] of the Matopo hills, as a curse which was due to the presence of the white man in the country. The visitations referred to were in the first place a drought of abnormal length and severity;

secondly, the appearance of enormous swarms of locusts, which did not seem to have troubled the country much in the previous quarter of a century; and, thirdly, the hitherto unknown *rinderpest* which destroyed whole herds of cattle.[62]

Among the Valley Korekore, each hamlet belongs to a land shrine neighbourhood. Each neighbourhood comprises several hamlets whose members are supposed to gather each year at a land shrine at the beginning of the first rains, in order to make an offering of beer to the territorial spirit in whose 'spirit province' their gardens are situated. The shrine usually consists of a baobab tree. Among the Eastern Korekore, it can also be a *mutuwa* tree (*Kirkia acuminata*) or a rock.[63] In the case of the last-mentioned, an enclosure (*rushanga*) is made out of poles around the tree and inside it are kept pots (*hupo*) for all territorial spirits who ever possessed a medium in the region, as well as for past chiefs whose spirits have been 'brought home'.[64] Participants in the ritual squat before the shrine and clap their hands in unison as is customary in saluting a chief, while the master of ceremonies pours a libation into the pots. Each of the spirits they represent are admonished to give the gardens a good supply of rain, to keep the pests away, and thus to provide the people with a good harvest, etc.:

> Kagoro, are you there? Don't you see us? Do you want to leave us here with hunger? Remember us so that you may keep us this year. We are dying of hunger. If the rain does not come we shall not cook for you again. What wrongs have we done you? Are you still outside, Kagoro? Here is your beer, drink it all this night![65]

Garbett adds the important observation that the tree shrine ought to be fairly near a pool or a stream filled by the first rains, and that the beer used in the ritual must be prepared with water collected from the first rains of the season.[66] In chapter three it will be seen how much these injunctions are bound up with Shona mythological lore. Bourdillon found that in the chiefdom of Chiruja no such rites were performed during the severe drought of 1970, in some cases the excuse being made that the spirit medium was sick, and he goes on to say:

> It appears that the ultimate and desperate measures with which the people used to meet the threat of starvation after a severe

> drought have become obsolete since dams have been built and boreholes drilled which safeguard the water supply, since meal has become available in the stores, and since it has become known that, should the situation become intolerably severe, the government would organize some famine relief.[67]

While there is undoubtedly an inherent trend in tribal society towards discarding costly rites, or at least to play a wait-and-see game with their territorial spirits, there remains a residue of readiness to submit to the customary rites when it comes to the worst. Thus Bourdillon observed that in Chiruja during the following season, when the crop was promising but threatened by an invasion of corn-crickets, three mediums were approached independently by small parties from the areas around their homesteads who sought their help. He notes that at the time people said that these rites used to work in the old days, but did not work properly nowadays. Later, however, when the threat had passed over, they said that they had been to their spirit mediums and that for this reason their crops had been saved. 'Thus', he says, 'the validity of the [spirit] cult was verified in the minds of the people, who, no doubt, would have been able to explain away any possible failure of the rites in terms of some breach of taboo or having the wrong chief'.[68]

Abraham notes that there are Shona chiefdoms in which the spirit cult no longer exists or is steadily diminishing.[69] Among the Karanga, according to Weinrich, 'the office of spirit medium carries prestige... but is inferior to that of a chief'. It seems to her that 'those who for structural reasons are disqualified from becoming a chief tend to fill this ritual office'.[70] However, Abraham makes the interesting claim that the cults of Mwari and of Chaminuka, which latter he regards as an off-shoot of the first, functioned as conservative 'pressure-groups' in defence of the political *status quo*, and that in this role they displayed a remarkable ability to adjust themselves to new situations in terms of political realism.

Thus, when during the 17th century the Portuguese installed rulers of their own liking, regardless of whether these had the right of succession, the spirit mediums found ways to oust such alien-installed chiefs in favour of their own nominees.[71] Garbett tells of a similar incident which happened in recent times. When a spirit medium was told that the administration refused the appointment of the candidate whom he had nominated during a previous seance,

he declared that he could not take back the title once he had given it. The people would have to solve the difficulty themselves. He warned them, however, that he would not recognize the White-appointed chief and threatened that any man who took the title without his permission would soon die, killed by the ancestral wrath.[72]
In a different context, the Protestant missionaries mentioned earlier on, found that their attack on the spirit cult was interpreted as an attack on the political *status quo*. Their first Sunday service in 1883, at Chief Zimuto's village, was attended by only one person and almost drowned out by the noise from a ceremony nearby, whose central focus was ten possessed spirit mediums.[73]
Before looking closer at their alleged possession, it should be noted that the spirit mediums' hold over the people is by no means universal, neither is it equally strong at all times. In an article Bourdillon wrote to discount the view which had been expressed in the Rhodesian press at the time, namely that the Shona had absolute faith in their spirit mediums, he maintains, that in fact, 'tribesmen normally accept a medium's directives only if they have sympathy for what the medium says'.[74] However, it appears that the people's faith in the spirit mediums' extraordinary powers tends to be more resistant to opposing pressures than is their esteem for the powers believed to be inherent in the office of chieftainship. It would seem that the reason for this tendency lies in the fact that, as spokesmen of the original 'owners of the land', the mediums themselves are seen as more immediate representations of those powers which in the Shona's experience are least controllable. This spokesmanship finds its typical expression in the medium's alleged possession by the territorial spirit.

Spirit mediums' possession

After the burial of 'owners of the land', their spirit is believed to enter into maneless lions (*mhondoro*),[75] roaming through the wild bush. For this reason, when a chief is buried, care is taken to leave an open conduit between his face and the surface of the grave, through which his spirit will be able to escape. After a minimum of five years' roaming in the wilds the spirits are said to seek out a suitable human host through whom they can speak to their people, but it appears to be largely a matter of chance whether or not a particular spirit 'comes out' in a medium.[76] Persons believed to be called to become mediums usually display stereotyped symptoms of illness, such as persistent headaches, lack of appetite, and recurring fits. They also tend to run into the bush in what seems to be a trance-like state, where they may roam around for days, eating wild berries and raw meat. If the illness persists and is interpreted by a diviner-healer as indicative of a territorial spirit's wish to use the person concerned as his medium, steps are taken to test the authenticity of the calling.

In some Shona country, territorial spirits are believed to choose as their mediums people from far away places, who are complete strangers to the territory in which they were the 'owners of the land' during their lifetime. In this way they are believed to demonstrate their persistent power, for the strangers whom they have chosen as their mediums are expected to possess exact knowledge about their genealogy, and about its relationship to that of other territorial spirits. They must also be sure about the boundaries of their 'spirit provinces' and display a wealth of esoteric knowledge about tribal history which is not generally known even among the elders. However, in fact, most mediums are found to have been born in or near the chiefdom of the territorial spirit who is believed to have chosen them as his medium.[77] Thus potential mediums have had the opportunity to acquire their special knowledge before their calling, usually by having been associated with an operating meduim. Besides, the required tests of persons claiming to be possessed by a territorial spirit are not always carried out, or are merely performed in a casual manner.

Nevertheless, 'the spirits are believed to control all the affairs of their chiefdoms and to communicate directly with their people when they take possession of their mediums'.[78] Possession usually takes place in the early hours of the morning and is heralded by more or less violent jerks of the medium's body and by his emitting deep roars and grunts. People show their respect to the possessed medium through removing their hats and shoes. Men clap their hands and women shrill, as they do for chiefs and during rituals in honour of their ancestral spirits. Questions are directed to the possessed medium through his principal acolyte – sometimes referred to as his 'interpreter' (*muturikiri*) – who conveys the territorial spirit's answers to the questioning party.

Questions are not always answered immediately. Sometimes the medium orders beer to be brewed and then leaves to consult with other spirit mediums before giving his answer.[79] Also Bourdillon's material strongly suggests that the mediums' falling possessed is to a large degree under their control. They manipulate it skilfully during the vagaries of the often long drawn-out succession disputes in their chieftaincies, letting it occur and holding it back as it suits them. Mediums can be used by families as levers in their attempts to bring their own candidate for the chiefly office into power; or, putting it differently, ambitious people may be tempted to pose as mediums in order to promote the cause of a particular family in its struggle for chieftainship.[80]

It has been observed that among the Shona '*today* contenders can intrigue and scheme with no respect for popular support', because 'both spirit mediums and people acknowledge that ultimately it is the government that appoints [chiefs]'.[81] However, it is reasonable to assume that even before the establishment of the European administration in Rhodesia, the spirit mediums exercised the same contriving role during succession disputes for the chieftaincy as today. Although it is true that in the heat of the battle particular spirit mediums may be publicly decried as liars, and, as a consequence, the whole corporation's credibility may reach a low ebb, nevertheless, at least among the rural population it still retains considerable potential for a come-back during times of crisis.

If the rains fail, or if pests or wide-spread disease occurs, people continue to approach the spirit mediums in large numbers in order to seek their advice. Falling possessed, the mediums then inform the petitioners in what ways they have offended the territorial spirits and what they have to do in order to make amends. Furthermore, as I have already indicated, the spirit mediums have proved themselves capable of playing yet another important role in the more recent past of the Shona people, i.e. that of inspiring and organizing popular resistance on a far-reaching, supra-tribal scale.

The Shona Rebellion

Before the outbreak of the Shona Rebellion in 1896, White settlers and administrators were unanimous in their opinion that the traditional disunity characteristic of the Shona paramountcies, and the continuous squabbles between chiefly lineages in most of the chiefdoms, would make a concerted rising a virtual impossibility. When the Amandebele launched their short-lived rebellion in January 1896, Native Commissioner 'Wiri' Edwards in the Hartley district was still convinced that the Shona would not join in with them, although he later recalled that there were 'some signs of strangeness' among the people in his district.

Among other things, there was talk about a 'little bird from Mwari' having arrived in the chiefdom.[82] This euphemism was employed to conceal the fact that messengers had arrived in this western Shona chiefdom from the Mwari shrines in the Matopo Hills. They urged the Shona chiefs to join the rising of the Amandebele, their traditional enemies, against the foreign oppressors. The messengers were believed to be able to work miracles, such as 'doctoring' the rivers so that the White men and their horses would burn up when entering the water. They told the Shona warriors that the bullets fired from the white men's guns would be as harmless as water.[83]

Outside the area which was traditionally affiliated with the Mwari cult, in central and eastern Shonaland, the spirit mediums at the heart of the rising were representatives of the territorial spirits Chaminuka and Nehanda.[84] Ranger writes that, as one would expect, 'no one senior medium enjoyed a determining influence over the whole area of the Shona Rising, nor did the mediums as a whole unanimously agree to support it'. But he goes on to say: 'The influence of the senior mediums was even in normal times more extensive than that of any secular authority, and could bring together for ritual purposes rival, even warring paramounts'.[85]

It is interesting to note that, at the time of the rising, there was, in fact, no recognized Chaminuka medium, though there were several imposters. The claim has been made that even Mkwati, who fomented the rebellion from the Mwari shrine Ntaba-zika-Mambo in the Matopo Hills, was not a 'Child of Molimo', i.e. a descendant of the prestigious Rozvi dynasty, but that he belonged to one of the 'slave' (*holi*) tribes.[86] The fact is, however, that the Mwari cult and the hierarchies of mediums elsewhere, formed the conceptual background out of which there could 'emerge charismatic figures, whose . . . power of personality was capable of throwing relationships within the cults into a new pattern, and who were able to break away from the established interrelationship with political authority'.[87]

Thus one of the two outstanding organisers of the rising in western and central Shonaland, Kaguvi, was – in the words of the local Native Commissioner – before the rebellion 'of no more importance than any other common or barn door *mhondoro*', though it was believed that he had special powers to locate game and to enable people to hunt it down. In contrast to his insignificance, the other organiser, Nehanda, was 'by far the most important wizard in Shonaland' and received tribute from all the chiefs who took part in the rebellion.[88] However, after having made contact with Mkwati, Kaguvi claimed to be the natural brother of Chaminuka, and a personal representative of Mwari, known under the title Mulenga or Murenga.[89]

From their headquarters, their 'Birds of Mwari' were sent out to summon representatives of the various chiefdoms to them, under the pretext of supplying them with 'medicine' against the locusts. While in trance, possessed by Murenga, Kaguvi gave 'orders for the destruction of anybody obnoxious to him' and told people that all Whites had to die this year. Any booty taken from looted farms had to be sent to the mediums. They hoarded it and used it as gifts to spur the courage of distinguished fighters.[90]

The Shona Rebellion lasted well into the latter half of 1897, and even after it had been ruthlessly put down, the surviving chiefs and mediums remained in a ready mood to fight the White men. There was talk about an imminent second rising during the Boer War, and, in 1903 when the hut tax was increased, 'unusual activity amongst the witchdoctors' was noticed and coincided with the 'coming out' of a new medium of Chaminuka, in the Hartley district. This was the message:

> I am Chaminuka. I know everything. I am all-powerful. I caused the downfall of the Barozwi and the Matabele and I will cause the white man to leave the country. Nothing is impossible to me. Follow my instructions. Makombe is coming soon. Why, he is one of my children. He will not stay away.[91]

It is small wonder that the Rhodesian nationalist leaders, who had been for a long time wary of appeals to the tribal past, eventually discovered that one of the most effective means to shake the Shona out of the lethargy which had gripped them by the 1950s, was to appeal to the great supra-tribal territorial spirits and to the spirit mediums who had been at the heart of the rebellion. They began to see the latter as forerunners of their own attempt to create a new, supra-tribal nation, and they realized, as Ranger notes, that 'the rural masses, while profoundly steeped in the Shona tradition, would follow urban nationalist leaders against their own chiefs in the very name of that tradition'.[92]

NOTES

1. Latham (1974), p. 96. Cf. also Hamutyinei and Plangger (1974), pp. 70f. 87.
2. Muzukuru = agnatic descendant of the women born in one's own agnatic group. Therefore also 'grandchild'.
3. Parkinson (1972), p. 64.
4. Fowle (1973), p. 29.
5. Bourdillon (1972a), p. 87. In the past, the corpse of a deceased Shona chief used to remain on a raised platform inside a hut for several months. The walls of the hut were carefully kept in good repair to prevent the deceased's spirit from escaping, and for the same reason pots were placed underneath the platform, to collect the fluids and maggots emitting from the corpse. Cf. Edwards (1929) p. 38. For greater detail on these practices cf. Frobenius (1930), p. 134f.
6. Fynes-Clinton (1972), p. 63.
7. Katsukunya (1971), p. 77f.
8. Latham (1972), pp. 78, 80; cf. also Bourdillon (1972a), p. 88.
9. Latham (1972), p. 80.
10. Bourdillon (1972a), p. 285. There is an interesting parallel among the Swazi. Cf. Kuper (1969), p. 627f: The Swazi king 'must be identified with the many and at times conflicting forces in the cosmos'. One way in which this is symbolically done during the annual *Ncwala* national festivities is through his right to break the rules that constrain ordinary men (by being allowed to wed women from all clans in the nation, including his own). The 'Hand Song', which is sung during the *Ncwala*, says of the king that 'he sleeps with his sister': cf. Kuper (1947), p. 203.
11. Bourdillon (1972a), *loc. cit.*
12. Hatendi (1973), p. 136.
13. Howman, (1959), p. 40.
14. An alternative meaning of *simba* is authority: cf. Hannan (1974). Note the parallel variety of meanings of the same term among the Temne of Sierra Leone: 'The term they use to designate "strength" is *ofoso*, which as in our everyday use of the words "power" or "strength" can mean several different sorts of effectiveness – physical strength or power, political or social power, and the sort which, because we do not believe in it, we usually qualify as "supernatural" or "magical" '. Cf. Littlejohn (1973), p. 288.
15. Weinrich (1965), p. 14-16.

16. Bullock (1928), p. 70.
17. Incorporated into the myths which recount the way in which the various spirit provinces came into being are also accounts of how each man who is now believed to be a spirit medium ate special 'medicines' before he died. By virtue of their power his spirit was able to emerge from his grave and to seek out a special lion (*mhondoro*) and eventually a human medium through whom he could speak. Cf. Garbett (1963), p. 208.
18. *Ibid.*, p. 44f. Cf. also Garbett (1969), p. 107-13.
19. Bourdillon (1976), p. 299f.
20. Cf. Daneel (1970), p. 57.
21. Cf. Daneel (1970), p. 69f; Daneel (1971), p. 89f.
22. Cf. Burke (1969).
23. Daneel (1971), p. 80f.
24. Ranger (1976), *passim*. The cult seems to have spread to Vendaland only after the Amandebele invasion.
25. Bourdillon (1976), p. 326.
26. Daneel (1971), p. 81. On p. 89 he notes that during the big annual rain ceremony which is held at Matonjeni, delegations petition Mwari for rain, health and fertility. These requests are identical with those made to territorial spirits elsewhere.
27. *Op. cit.*, p. 321. My emphasis.
28. Daneel (1971, p. 84f.) unwittingly concurs with this view.
29. Bourdillon (1972a), p. 173.
30. Marr (1962), p. 82.
31. Garbett (1963), p. 54f.
32. Bourdillon (1972a), p. 131.
33. Latham (1974), p. 106f.
34. Dos Santos (1609): cf. Welch (1950), p. 154f.
35. Cf., e.g., Gluckman (1954), p. 30f. & *passim;* (1955), p. 110-28.
36. Welch (1950), *loc. cit.*
37. Zvogbo (1973), p. 63f.
38. Marr (1962), p. 82.
39. Bourdillon (1972b), p. 116.
40. Beach (1973), p. 38. Stayt (1931), p. 43 mentions a similar incident among the Venda, when chief Mphephu attempted to abandon his people's traditional attachment to the Mwari cult in favour of the famous rainmaker Majaji near Pietersburg in the Transvaal.
41. Cf. Garbett (1963), p. 250f.
42. Latham (1974), p. 97.
43. A curious exception would seem to be the arrangement, supported by mythological charter, between the chiefdoms of Charewa and Mtoko, where the ruling chiefs are said to appoint their own successors, although a spirit medium is also involved in the process: cf. Berlyn (1972), p. 56f.
44. I choose my examples from among the better documented 19th century events, though the basic tenor which is recognizable in them comes through in early Portuguese records as well.
45. Beach (1973), p. 34f.

46. Atkinson, (1973), pp. 85-8.
47. Peaden, (1973), p. 71.
48. Katsukunya (1971), p. 78.
49. Vambe (1972), p. 168.
50. Ranger (1967), p. 378, Cf. the Mandari expression: 'before the earth was destroyed', meaning 'before the slave-raiding started', which is used in dating events of the past: Buxton (1973), p. 150.
51. Garbett (1963), p. 196f. Cf. also Mbee (1965), p. 201: 'Now that we are independent, we can do as we wish. Our first desire is to make our country pure so that it will again produce sufficient food . . . But we are obstructed in our attempt to purify this country by the mission people who refuse to participate in the *mpefo* ritual and who continue to hoe on the days when this is forbidden because it would spoil the medicine'. (A Statement by the Mbugwe elders (Tanzania) during a meeting).
52. Ranger (1967), p. 353.
53. Garbett (1963), p. 157. Fifty years before him, H. Rolin had already remarked that 'les masses ont bien vite discerné où est la puissance effective; le prestige ancien des chefs est atteint: ce sont les commissaires indigènes ou les commissaires de district qui en héritent et qui deviennent les seules autorités véritables'. Cf. Rolin (1913), p. 219.
54. Cf. Weinrich (1971), p. 162. His indictment echoes that which Kenyatta had made against Kenyan chiefs in 1938: cf. his *Facing Mount Kenya* (1953 ed.), p. 225f.
55. Dombashawa is the name of the place where the meeting (*indaba*) was held. Alternative spelling: Domboshawa. Cf. *Rhodesia . . .* (1964).
56. Ranger (1968), p. 215. Cf. also the criticism of Holleman by H. J. Simon for his failure to take into account the rival claims of the Rhodesian nationalists, in his book entitled *Chief, Council and Commissioner* (1968): Simon (1971).
57. *Pearce Commission* (1972), p. 49.
58. Weinrich (1971), pp. 78-87.
59. Garbett (1963), p. 243. My emphasis.
60. Bourdillon (1972a), p. 24.
61. Ndebele term for Mwari.
62. Cf. Nobbs (1924), p. 41.
63. Garbett (1963), p. 182; Bourdillon (1972a), p. 142.
64. For the meaning of this term, cf. chapter three. In my short description of the ritual I include elements from both Bourdillon's (1972a) and Garbett's (1963) accounts. Cf. also Garbett (1969), p. 119f.
65. Garbett (1963), p. 194.
66. Garbett, *op. cit.*, p. 188f. For a description of similar rites arranged by the messengers of Mwari in their respective districts and at the cult's main shrine cf. Daneel (1970), p. 58f.
67. Bourdillon (1972a), p. 166. At the time, the only land shrines existing in Chiruja were in the Western part of the chiefdom. The reputation of the spirit mediums was at a low ebb, because of the senior medium's unpopular handling of the latest chiefly election.
68. Bourdillon (1972a), p. 168.

69. Abraham (1966), p. 29.
70. Weinrich (1971), p. 113.
71. Abraham (1966), pp. 38; 43.
72. Garbett (1963), p. 248.
73. Beach (1973), p. 36f.
74. Bourdillon (1974), p. 30.
75. 'Ordinary' lions are called *shumba*. Cf. also above, p. 34, n. 17.
76. Bourdillon (1972a), p. 53. In the chiefdoms where he did his research, a founding chief and his sister will always come out as territorial spirits.
77. This is, e.g., the case with the 18 spirit mediums of the two chiefdoms described in Bourdillon (1972a), p. 59. Cf. also Garbett (1963), p. 238f.
78. Bourdillon (1972a), p. 43. I will discuss the 'reality' of possession in chapter four.
79. Garbett (1963), p. 232. In his own case, Garbett says, all but one medium refused to speak to him unless they were possessed. On his arrival at a hamlet where a medium resided, the acolyte would inform the medium of Garbett's presence and then go to bed, 'until he was awakened by the roars and screams which signified that the spirit had arrived': *loc. cit.*
80. Cf. Bourdillon (1972a), pp. 52-7; 65; 170; (1976), p. 127f.
81. Cf. Bourdillon (1972a), p. 113f. My emphasis.
82. Ranger (1967), p. 192.
83. Ranger (1966), p. 122.
84. The north-eastern Shona, Korekore, and Tavara did not join the rebellion.
85. Ranger (1967), p. 207.
86. Cockcroft (1972), p. 85. How well the cult succeeded in keeping its activities secret, and how little the Whites at the time understood its functioning is documented by the macabre stories about the shooting of a 'Mlimo' in his cave in Kalanga territory, quite removed from the area of the rebellion. At the time it was celebrated as a daring episode, aimed at forestalling a rising in that strategically important region. Cf. Baden-Powell (1897), 'p. 81f; Nobbs (1924) p. 43.
87. Ranger (1967), p. 213.
88. *Ibid.*, p. 215.
89. Significantly, Rhodesian Black nationalists refer to their armed struggle against the Smith regime as *Chimurenga*. Cf. *The Daily Telegraph*, May 17th, (1976), p. 4.
90. Ranger (1967), pp. 210, 218, 220; cf. also Vambe (1972), pp. 118; 123.
91. Ranger (1970), p. 10. Makombe, a Tavara chief, at the time fought the Portuguese administration in Mozambique.
92. Ranger (1967), p. 382. Shortly after the conclusion of the Rhodesian internal settlement in 1978, one of the many new Black newspapers in the country, *The Rise*, carried the full story of Nehanda's leading role in the Shona Rebellion. Cf. *The Star*, 21. 3. 1978, p. 26.

CHAPTER 3

'GUARDIANS' OF THE FAMILY: THE ANCESTRAL SPIRITS

The role of the territorial spirit in relation to the tribe, or a large segment of it, is fulfilled, on the level of the extended family, by the *midzimu*,[1] the family's ancestral spirits. Given the fact that in the Bantu languages each word is assigned to a specific noun class indicating the field of notions or phenomena to which it belongs, it makes sense to go into the etymology of *mudzimu*. The *Standard Shona Dictionary*[2] defines *mudzimu* as belonging to the first noun class, which is used for persons. However, it states that the plural is *midzimu* or *vadzimu*,[3] in this order. The plural form *midzimu* belongs to the third noun class, and is used as a prefix for names of trees, some things characterized by length, and atmospheric phenomena.[4]
By merely glancing over words listed in this noun class one encounters innumerable kinds of 'medicines' and charms, and such atmospheric phenomena as *mubvumbi* (persistent drizzle; wet mist), *muchenganherera* (general rain), *mwando* (moisture, moist wind), *mutsiwemvura* (rainbow), *mwedzi* (moon), and *mwaranzi* (sunbeam). Noting that *mweya* (3rd class: air, breath, vapour, spirit, soul, . . .) and *mudzimu*, bear a close etymological relationship to words used to denote similar concepts in a large number of languages throughout Bantu-speaking Africa, Murphree writes:

> *Mudzimu* is in a class of Shona nouns which denote objects which are light, move, change, grow, produce, or, in general, which contain some principle of life and production, a notion intimately connected with that of 'power of growing up' like a tree.[5]

Roaming in the 'bush'

Bourdillon has made an extensive survey of Shona ideas concerning the physical nature of their ancestral spirits.[6] Naturally, these ideas are very vague. People use the term *midzimu* primarily as a symbolic concept; they are, therefore, not really interested in the physical shape of their ancestral spirits. This is precisely the reason for assigning them to the third noun class: they belong to nature and have power over it, for 'a ghost is not just a person who has died; it is a being of quite a different order from the living . . . Ghosts are left by people, but they are not people'. [7] Their absorption into the realm of nature is emphasized through the way in which people visualize the post-mortem existence of an ancestral spirit. Some say that it leaves the grave of a newly buried person in the shape of a variety of animals or insects, others maintain that at a person's death it hovers about the corpse as a shadow[8] and then vanishes 'like air' and may be anywhere thereafter.

There is, naturally, even more vagueness about the question when and from whence a person's *mudzimu* is supposed to come about in him, but there is no need to go into this here.[9] What is important is that every grown-up person[10] is attributed to releasing an ancestral spirit upon his death, and that this *mudzimu* is conceptually linked with the realm of nature. It is thus symbolically associated with a field of experience containing aspects which are both frightening (e.g. lightning, hail, drought, animals harmful to man, livestock, and crops) and benevolent (e.g. rain, forest fruit, the harvest, domesticated and 'friendly' animals).

Looking at the fate of a chief's *mudzimu* upon his death also throws light upon the fact that ancestral spirits are conceptual constructs. In as much as he was a chief, his spirit is supposed to turn into a territorial spirit which will look after the interests of the whole tribe. However, since he was also a member of a lineage, his spirit has to be 'brought home'[11] by his family, and it is believed to fulfill thenceforward the same role of ancestral spirit vis-à-vis his descendants as that of any commoner's.

Before the final funeral ceremony has been held, during which the ancestral spirit of a recently deceased is 'brought home' to its homestead, it is imagined to be vaguely 'outside', roaming about in the veld or floating somewhere in the air.[12] It may also find temporary quarters in some large tree. Such a newly released ancestral spirit may be addressed and honoured anywhere in the bush, and rites in its honour are often performed in the fields of the senior agnate of the deceased's lineage. However, as it is still a novice, it is believed to be able to cause only a mild sickness among one of its descendants if it wants to be honoured by them.[13] Logically, the 'bringing-home' ceremony begins in the bush[14] and moves from there to the deceased's former homestead. When a person has died, any valueless possessions he has left behind are deposited somewhere in the veld.[15]

The notion that the bush or the open veld is the temporary abode of the ancestral spirits is only one in a chain of variations on the same symbolic theme, which the Shona share with many other peoples in southern and central Africa. For instance, it is tied up with the idea that the first people emerged from an abyss, out of which all life comes and to which their spirits must again return.[16] The same idea is expressed in tales which report that in the past the voice of the widely known territorial spirits, Mwari and Chaminuka, could be heard from rocks and from trees, and that people disappeared into rocks at the end of their lives.[17] The Shona also possess a variety of myths surrounding the theme of the *dziva*, a mythological pool, on the bottom of which there are exact duplicates of the customary Shona homesteads peopled by the ancestral spirits, herds of cattle, pastures, etc.[18] In spite of the theme's many variations, its basic message is always the same. It gives expression to the experiential fact that at his death man enters into the unknown.

Until it has been found out by divination why a particular person passed away, death vaguely threatens the whole family group of which he was a member, and it may soon strike again. It is particularly feared that the deceased might have died harbouring an unresolved grudge against a surviving member of the family or against an outsider, in which case it is feared that his ancestral spirit will afflict the guilty party in the form of an avenging spirit (*ngozi*).[19] Besides, as it is often suspected that death occurred as a result of witchcraft, there is almost always reason to assume that the deceased left this life in a state of 'hotness'.

'Cooling' the ancestral spirit

The similarity between the Shona notions of 'hotness' and 'coolness' of spirits and the English idiom has been pointed out by different writers.[20] However, it can be justifiably argued that the Shona's repeated experience of the devastating results of drought lends particular expression to their consideration of 'coolness' as the desirable normality both in nature and in a person's state of mind. As will be seen later,[21] for them 'coolness' is also cognitively linked with the whiteness of rainwater and of meal.

According to Shona informants, the purpose of 'cooling' the spirit is 'to help the deceased to settle down in his new mode of existence as a member of . . . the spirit world, without severing any of the bonds with kin and friends left among the living'. [22] Funerals may, therefore, only take place in the 'cool' of the morning or evening; water or porridge is poured over the grave after the burial, and a shading branch is planted on top of it. In addition, the spirit himself might be bidden farewell and asked to care for those who remain. During the ensuing ceremonies, roasted meat is eaten unsalted; and before returning to their homesteads, all mourners have to wash their faces, arms, legs and shoes.[23] Ashes remaining from any fires lit in connection with the funeral ceremony are deposited in the bush with the declared intention thereby to 'cool' the spirit of the deceased.[24]

The need to 'cool' the new ancestral spirit is most dramatically expressed in the fact that throughout the burial the deceased's classificatory daughters-in-law have to lie down in his honour a few paces away from the grave and feign sleep. It is quite obvious that through this gesture the 'coolness' of peace and security ruling the homestead is symbolically emphasized. This visible dramatization of the desired state of normality is indeed very necessary, because immediately after the burial the first informal discussions take place with a view to determining the successor to the rights of the deceased over his property, including his women and children.

Until the inheritance ceremony is finally held, there is bound to be mounting tension between rival parties. However, it is particularly feared that one of the younger widows might 'burn the grave'[25] by indulging in sexual intercourse from which they must abstain until the inheritance ceremony has taken place. No close relative may even touch any of the deceased's possessions until after that ceremony, and in order to 'cool' them they are sprinkled with specially 'medicated' water.[26] Utensils which the deceased has handled daily, such as plates and spoons, are either buried with him or placed upon the grave after the burial, 'to ensure that [his] spirit . . . does not trouble the other members of the family, thereby accusing them of neglect'.[27]

Authority and ownership

The inheritance ceremony can only be held after the spirit of the recently deceased has been 'brought home'[28] and settled in the ranks of the ancestral spirits. These are thought to be more predictably beneficient, though they too are held 'in a mingling of fear and confidence',[29] because

> they remain in the world of the unknown, and the living are always a little afraid about how they will use the super-human powers they are believed to possess. So they must be pacified and cajoled into doing their duty, and should be given no excuse to be displeased.[30]

The Shona share the ambiguous attitude towards their departed ancestors with many other societies in Africa. For this reason, Igor Kopytoff's suggestions some years ago that in Africa 'an elder's social role does not radically change when he crosses the line dividing the living and the dead',[31] was criticized by J. L. Brain, who gave a number of reasons why Kopytoff's simple equation of ancestorship and eldership cannot be upheld. Drawing his evidence from various African societies, Brain pointed out that for the Luhya of Western Kenya a *mzimu* is not an ancestral spirit, but rather the power used by the spirits to punish their descendants. For him, therefore, 'the key word here is authority. One who has authority is in contact with power'. [32] However, Brain fails to show convincingly that which lies at the heart of the ancestors' power over their descendants. He sees it in the fact that they are not only able to observe their descendants' actions, but that they are aware even of their innermost thoughts.[33] I, however, believe it is important to note that, at least for the Shona, authority is most notably and inextricably linked with rights over property and ownership.

The link between authority and ownership has already been encountered in the first chapter during my discussion of the notions that go with the term 'owner of the land'. These notions have been seen epitomized in the recalcitrant subject's question to his chief: 'Where is the land?'. The same linkage can also be observed on the level of the family. Bourdillon notes that 'perhaps the most important relationship to the Shona' is that between a father and his child.[34] The father has absolute authority over his children and complete responsibility for them. A child must 'fear' (*kutya*) his father and at all times display an attitude of respect towards him, reverently clapping his hands when he meets him. He is not allowed to take any liberties with his father's property.

A special relationship also exists between a brother and that sister (*cipanda*) whose bridewealth has been, or will be, used to obtain a wife for him. She holds a particular position of authority over her brother's children, since it is through her marriage that their family was able to come into existence:

> She is 'owner' (*vamwene*) to her brother's wife (the 'wife of her cattle') and 'controller of the wife' (*samukadzi*) when she officiates in matters of family dispute or in connexion with the distribution of the family estate, when she exercises undisputed authority... The relations between wife-receiving males and wife-providing females are characterized by the assumption that any sister or father's sister has provided a wife to her family and is therefore able to exercise such authority.[35]

On the other hand, as in many other African societies, the mother's brother/sister's son (*sekuru/muzukuru*) relationship, is also marked among the Shona by intimacy and cordiality, and the sister's son is free to avail himself of his mother's brother's property, without requiring permission to do so. He shares that particular kind of authority which a sister may exercise over her brother and his family and enjoys a degree of freedom in the homesteads of his mother's brothers which he would never be given in those belonging to the senior members of his agnatic group. Because it was the mothers of the sisters' sons whose marriage secured the cattle which made the marriage of their brothers possible, a sister's son may tell his younger mother's brothers that they are all 'his children'. Their mother is his 'wife' (*muramu*): 'Has she not been married with my (i.e. my mother's) cattle?'[36] For this reason, the term 'controller of the wife' (*samukadzi*) and the functions pertaining to it, apply to the uterine nephew as well as to a sister or father's sister, and on occasions such as the distribution of a deceased's estate both the deceased's father's sister or sister, and his uterine nephew jointly officiate.

The same principle also dominates the relationship between the respective lineages, lumped together into 'wife-givers' (*vatezwara*) and 'wife-receivers' (*vakuwasha*). As will be seen, the former can rely on a number of services which the latter are bound to render them. However it has been stated that, in a sense, the 'wife-receivers' only 'hire' the reproductive capacity of a woman given to them by the 'wife-givers'.[37] Their payment of cattle as bridewealth imposes a strict obligation on the 'wife-givers' to provide the 'wife-receivers' with children. Should the 'hired' woman prove barren, or unwilling to fit herself into her husband's family, the 'wife-givers' have to give them an acceptable substitute or return the full bridewealth.

When the wife dies before bearing many children the 'wife-givers' have to provide a substitute, or return a proportionate part of the bridewealth to the 'wife-receivers'. The inextricable link between the transfer of cattle and the right to parenthood is further emphasized by the fact that after the payment of bridewealth to the 'wife-givers', the 'wife-receiver' is the recognized genitor of any child borne by his wife, no matter who the child's physical father really is. This legal point has its natural corollary in the institution of the levirate which comes into action upon the death of the 'hired' woman's husband. It is against this background that the Shona expression, 'Cattle are like people',[38] assumes its full relevance and as such needs no explanation.

How important it is to see this link between ownership and authority in order to grasp the power attributed to the ancestral spirits, becomes clear from the following statement by Aquina:

> (1) Ancestors are important not because they have been good and great persons, but simply because they are physically related to the living. (2) Even a morally bad person becomes an ancestor, and his power is not less because he was little liked during his lifetime. (3) The power of an ancestor-spirit, rather, depends on the social status he occupied in life. (4) Thus the *mudzimu* of a chief has more power than the *mudzimu* of a commoner, because it controls not only the chiefly family, but also all the people living in the chiefdom.[39]

While sentences (1) and (2) speak Kopytoff's language, sentences (3) and (4) contain, albeit in a veiled manner, the proof of my aforegoing argument. It is possible though to misunderstand the last sentence in Aquina's statement. It is true that, as territorial spirit, a deceased chief can affect the life of large numbers of people in his former chiefdom, through the influence he is believed to have on the well-being of the crops. It would be wrong, however, to attribute to his ancestral spirit the ability to afflict each and every person in his former chiefdom individually. Garbett is more precise when he states that the ancestral spirits of a headman are believed to affect the well-being of both his agnatic and cognatic descendants in the hamlet (*chizwarwa*) over which he has authority.[40]

These hamlets are small in size, are often moved to new sites, and are always in danger of breaking up as each of the senior males living in them eventually desires to found his own hamlet. In this situation, it is obvious that the headman's deceased ancestor (*tateguru*) who founded the hamlet, is an ancestral spirit to virtually all those who live in it. As its founder, he 'owns' the hamlet's reproductive means: land, livestock, men, women, and children. As their 'owner', he is responsible for preserving and increasing them in number, and any interference with them provokes his wrath, whether it comes from within the hamlet or from without. He metes out the same harsh punishment – illness, death, and misfortune – to both 'insiders' and 'outsiders', as ancestral spirit to the former, and as avenging spirit (*ngozi*) to the latter.

It is my opinion that the ancestral spirits are conceptualizations by which the Shona fuse two fundamental experiences which they come across in their daily lives, viz. the inescapable facts of our human existence – death, illness, and misfortune – and the authority and power held by those enjoying rights of ownership. To make this assumption does not mean to succumb to 'armchair anthropology', because both are equally real experiences in the lives of every Shona, and their characteristics are recognizable in the very way in which the spirits are conceptualized.

The elusive character of what in other societies outside sub-Saharan Africa may be termed 'fate', is expressed by the Shona in the conceived unpredictability of their spirits' actions, and results in the fear man will never quite succeed in satisfying their wishes. However, another, and equally important experience also finds expression, namely that holders of authority may treat harshly those who are under their power, but that submitting to them, in the long term, brings more advantages than attempting to oppose them; and that the period of subservience turns ultimately for everyone into one of dominance over others – for men more definitely than for women. Every Shona learns this lesson from first-hand experience while he is occupying his well-defined place in the 'tribal system'. It accounts for his belief that man can accommodate himself to the erratic powers which impinge on him and his world.

Perhaps linking the Shona's attitude to ownership and authority with their concept of spirits, smacks of Frazerian manipulation. However, it has been expressly stated by Bourdillon that

> they [i.e. the Eastern Korekore] also believe that very old persons begin to acquire the powers of spirits before they die. Dead persons maintain as spirits their positions in the social order, and they can still be approached as relatives and as friends;[41]

However I would add, as 'friends' and 'guardians' who inspire both fear and confidence in those who feel dependent on them. Anything that can be interpreted as failure on the part of their descendants to keep alive their memory is liable to provoke the ancestral spirits' vengeful action against any member of their kin, though children are believed to be their primary target. Most often they are thought to afflict their descendants with a mild illness, but they may also cause other misfortunes to befall the family's livestock and property.

Maternal ancestral spirits

It is important to note that the spirits are always thought to threaten the integrity of the family's property, 'property' taken in a wide sense: its homesteads; its cattle, fields and people – and the fertility of all these; and its 'luck' or good fortune. In this connection it must be mentioned that maternal ancestral spirits are particularly feared if divination reveals them as the cause of the affliction. The reason for this is twofold: firstly,

> because the mother carried her offspring in her womb, underwent pregnancy and the pains of labour, and nourished and cared for them while they were helpless children, at the death of the mother her spirit demands to be well remembered and honoured. If her children fail in their duties, the spirit is believed to have the same absolute power over them that she had as a mother during their infancy.[42]

I shall give an example below of alleged action by a maternal avenging spirit against her own son which illustrates this point. However, as will be seen, maternal ancestral spirits are particularly feared by a married woman's husband and his kin. This forms the background of the second part of Bourdillon's statement:

> Maternal spirit elders are so important as custodians of a child's well-being that occasionally a child's father's family waives its legal rights over a sick child in order to allow it to live with its maternal kin who can more readily communicate with its mater-

> nal spirits . . . Should a woman have any problems over fertility or child-bearing, her maternal spirits are presumed to be responsible and must be appeased.[43]

This is more useful reasoning than Aquina's oblique reɟerence to matri-lineality when she writes that among the Shona 'the spirits of mothers are considered more dangerous than the spirits of fathers, a common feature in matrilineal societies'.[44] Holleman, in his standard work on Shona customary law,[45] also refers to 'matri-lines' and 'matri-lineal descent'. However, to speak of matrilineal descent in connection with the Shona is misleading and fails to render a satisfactory explanation of the phenomena for which the use of this concept is intended to account. These are far better understood if they are seen against the theoretical background which I have proposed in this chapter, viz. that ancestral spirits are always actively involved where people's authority and property are concerned.[46]

It is important to note that Holleman does, in fact, often speak of the 'matri-group', whose 'line of descent' comprises no more than three generations: it goes from the child's mother to her own mother, and from there to the mother's grandmother.[47] Of equal importance is his statement that children's knowledge that they have been born out of the 'same womb' and therefore share 'the blood of the mother' establishes a closer sense of cohesion between them than the knowledge that they have been sired by the same father.[48] A child's lineage membership entailing his acquisition of his father's name and, later, accession to rights and duties, as well as inheritance of property in the lineage – all this is determined by his having the 'same blood' as his father. A son's having the 'blood of another' (man) may disqualify him from succeeding to the name and dignity of his legal father, even if he has been accorded the latter's clan and sub-clan name.[49]

I have mentioned above that, in a sense, the 'wife-receivers' only 'hire' the reproductive potential of a woman given to them by the 'wife-givers'.[50] There is, therefore, some justification in seeing married women as 'stranger-wives' living away from their kinsfolk, whose own relatives will dig their graves when they have died, because they did not belong to their husband's people. In a polygynous homestead each 'house' (*imba*) forms a completely independent domestic unit, at whose centre is the genetrix whom her husband has to 'give a womb'.[51]

With the help of the male head of the homestead each 'house' is self-supporting. A wife may, however, raise more food than is needed for her household and acquire property through selling the surplus. Her husband has no claim to such property, which forms part of what Holleman terms her 'matri-estate'. Not infrequently, women, whose husbands are unable to assist them, sell their own livestock in times of scarcity to buy food for their children without expecting their husbands to replace these animals when conditions have improved.[52] After her death, a wife's husband must show to the representative of her lineage 'every bit of property belonging to the woman ... and the *tezwara* ['wife-giver'] will take control of such property'. Rather than wishing to conceal anything belonging to his deceased wife, a husband will hand over more than is strictly necessary, because 'he fears his wife's spirit'.[53]

The most valuable asset of a married woman's matri-estate is, of course, her 'herd of the mother' (*dangga ramai*), i.e. the livestock she acquired during her life-time and which is, 'as a legal entity, completely separate from the estate of her husband and his patri-group'.[54] It consists of one of two types, depending on the manner in which the animals have been acquired. Those earned through the woman's manual labour (*dzamavoko* = 'of the hands') have already been mentioned. The second type comprises the 'cattle and goats pertaining to motherhood'.

Each time one of her daughters is married, a mother is to be given a head of cattle, the 'cow of motherhood' (*yombe youmai*), because 'she has cradled the wife in her womb'. Often she dedicates it to the spirit of her mother. Holleman makes the important note that commonly the 'cow of motherhood' is only handed to the mother after her daughter has given birth to children. More often she has to wait many years, and it is not uncommon for women to die before they have been given their animal(s).[55] This frequent delay is likely to be at least partly responsible for the 'mystical tension', the 'attitude of mutual avoidance' (*kuckana* = 'to fear one another') and the absence of all intimacy between the mother-in-law and her daughter's husband.[56]

The tension between the husband and his wife's mother is also at the bottom of the belief that her daughter becomes dangerous to both her parents, but particularly to her mother, as soon as she enters her first pregnancy. As the daughter herself enters motherhood, it is believed that she would 'break the back' of her parents if she were to meet them before a special ritual, *kusunggira*, has been performed. One of its main features is that her husband must provide her parents with a male and a female goat. While the male goat, given to the wife's father, is ritually slaughtered and eaten, the female goat is added to her mother's matri-estate.[57]

From the afore-going it is clear that the fearfulness of the wife's maternal ancestral spirits is closely linked with questions of authority and property, and particularly with the fact that 'wife-receivers' must always be aware that in a sense they have only 'hired' the woman's reproductive potential. The wife's mother's and mother's mother's ancestral spirits are directly associated with the procreative powers of their female descendants. If they are not remembered, or if the wife's husband unduly delays the payment of 'cattle and goats of motherhood', their ancestral spirits avenge themselves by 'stealing the granddaughter's menstrual blood', causing her to become barren.[58]

The livestock of a woman's matri-estate may be used to help her own blood-relatives, such as her full-brothers or sons, to obtain a wife if they need assistance.[59] However, one or two female animals should be kept available for dedication or sacrifice to her maternal ancestral spirits and to her own ancestral spirit after her death. Holleman points out that in comparatively few cases is a sizable herd of cattle to be found which has originated from the 'cows of motherhood'.[60] He further notes that nowadays grown-up sons of a deceased woman occasionally claim full control, and the benefit of their share, in their mother's estate, arguing that: 'our paternal family paid full bride price for our mother. Why should her family again receive cattle from her?'. According to him this argument

> stresses the fact that the members of the deceased's matri-group belong to different patrilineal units whose interests do not coincide. As members of their father's family and patri-lineage (*rudzi*), the deceased woman's sons represent the interests of their own paternal unit, as against those of their maternal relatives who belong to a different patrilineage.[61]

Members of a 'matri-group', therefore, lose contact with each other, because local groupings are based on paternal affiliations. This has a profound influence on the ultimate control and destination of the matri-estate.[62]

The so-called Shona 'matrilineality' provides no satisfactory explanation of the 'dominant importance' of 'the spiritual supervision of, and mystical intervention in, matters concerning the deceased's livestock' which is said to be a 'peculiar aspect' of the matri-estate. Holleman found his informants 'unable to explain why a woman's spirit should be more concerned about her estate than a man's spirit about his patri-estate'. [63] They must be thanked for their inability to respond to the investigator's insinuation which was contained in his question. It is clear from Holleman's own evidence that maternal ancestral spirits are not *more*, but *equally*, concerned about their estate, as the spirits of a patri-line. It is true that the Shona are more afraid of the revenge of a maternal ancestral spirit, but this reflects the fact that the coming into being and the continued existence of a matri-estate is less assured than that of a patri-estate. The precariousness of its existence finds expression in the expectation of a more fearful vengeance from the side of the ancestral spirits against those who fail in their duty to establish or to maintain their rightful property.

The fact that maternal ancestral spirits are particularly feared by a woman's husband and his kin reflects also the ambiguous feelings which the latter often experience vis-à-vis the wife as a 'stranger-woman'. As a member of another lineage, she has a bridge-function and can easily incur the suspicion of having divided loyalties; because with her own matri-estate she helps not only 'her blood' in her children whom she has borne for her husband, but also her own blood-brothers. Her kin, in turn, call her son, the *muzukuru*, 'a child that somehow belongs "inside" his mother's lineage, because it is of the same blood'.[64] For this reason, wives are far more often suspected of witchcraft than husbands, the most common event giving rise to an allegation being the death of a child during or shortly after confinement.[65]

When a woman sacrifices her 'cow of motherhood' which she has dedicated to her mother or mother's mother, she may summon home her own sons and daughters and others closely related to her by their having come from the same 'womb'.[66] Such rituals in honour of maternal spirits are, however, uncommon. All formal rituals in honour of the ancestral spirits involving a large attendance are directed towards the family's agnatic spirits. These should also always be remembered at any purely social gathering, the senior agnatic representative spilling some beer for them on to the ground and sometimes adding a pinch of tobacco as well. In practice, however, many Shona families engage in what might be termed a 'wait-and-see-game', putting off the performance of rituals and neglecting to make token offerings to their ancestral spirits, until positive proof that their ancestral spirits feel neglected is established through divination.[67]

However, the ancestral spirits' authoritarian role as 'guardians' is emphasized by the commonly held belief that they allow even witchcraft attacks against their living kin if they feel they have been neglected.[68] Any breach of the order of life the ancestors themselves had to observe in their own days, and which they afterwards handed on to their descendants, causes their spirits' displeasure and prompts them to take retributive action against their own kin.

Avenging spirits

The avenging spirits (*ngozi*), whom I have mentioned already, represent to the Shona a particularly fearful type of affliction by the ancestral spirits. The attacks of avenging spirits are said to be very sudden and extraordinarily harsh, involving, for instance, the death of several persons of a family group in quick succession, or the failure of ailing members of the group to respond to treatment. Gelfand[69] says there are four kinds of avenging spirit; however, he lists five altogether. According to him, an avenging spirit can be the spirit of one of the following categories of persons who left this world in an aggrieved state of mind: 1) the victim of murder; 2) a servant (slave) who was ill-treated and deprived of his rightful recompense; 3) one from whom goods were stolen or borrowed and never returned; 4) a husband or wife who was neglected by his or her partner; 5) a parent who was deeply hurt by his own child. However, Bourdillon has correctly pointed out that 'the concept of an avenging spirit refers not so much to a type of spirit as to its supposed action, namely death and destruction in the victim's family.'[70] This explains the fearfulness which the concept holds for the Shona.

In the light of the link which I have suggested exists between the authoritarian 'guardian'role of the ancestral spirits and the Shona concepts of ownership and authority, it is easy to understand why the measures attributed to avenging spirits should be so extraordinarily harsh, and, consequently, so much feared by the people. All of Gelfand's 'types' of avenging spirit derive from people who have been grievously harmed in their role as holders of property and authority. Not surprisingly, therefore, 'in all cases, the full appeasement of the angry spirit involves severe punishment, either by loss of wealth or by extreme humiliation'.[71]

People are able to cite only rare examples of the supposed avenging action of a *ngozi*.[72] For this reason, Gelfand often opens the discussion of his different 'types' of avenging spirit by saying: 'Let us suppose . . .'.[73] Nevertheless, although the information supplied by him is largely based on theoretical discussions with his informants with respect to the typical manifestations of the various 'kinds' of avenging spirit, it is valuable, because of its persistent emphasis on the spirit's unwillingness to desist from attacks on the wrongdoers until what is proper to him and his family group has been fully restored.

A family that has been grievously injured in its rights of ownership and authority may, however, also take steps to 'awaken' one of its deceased members, and ask him to avenge the wrongs done to them by acting against the offenders as an avenging spirit.[74] Another possibility is that the avenging spirit chooses a host and through him makes known his grievances. An authentic case described by Garbett[75] is instructive, because it arose over a situation not mentioned by Gelfand, though it has obviously to do with his third 'type' of avenging spirit. In this rather complicated case, when cattle demanded in payment of bridewealth were not forthcoming despite a series of sudden deaths, and the dispute started to affect an increasing number of people in the hamlet; a woman married to one of the offending parties became possessed by the avenging spirit. He threatened to continue to afflict his great-grandchildren belonging to the 'wife-receivers' unless the dispute was resolved.

In this case the avenging spirit's harmful action was aroused by the offending party's failure to pay in kind what was due to the avenging spirit's descendants and his attack concentrated on their resources of manpower and wealth. Gelfand writes of the Zezuru, that if the avenging spirit's grievance derives from his having been treated meanly and disrespectfully during his life, besides demanding compensation in kind he will also humiliate the offenders in public. Thus, for example, a son who had once beaten his mother during her lifetime was instructed to dedicate an ox to her. [76] Then, dressed in old sacks and pieces of blanket, he had to go begging in the neighbouring villages, until he had brought together a large enough quantity of millet from which beer could be brewed in his mother's honour. He himself did not participate in the feast for which the ox was slaughtered and during which the beer was drunk. According to the same source, similar punishments 'to the point' are meted out to those offending against other norms concerning respect for authority in Shona society.[77]

By now the reason for my having put 'guardians' in the title of this chapter between inverted commas should be clear. I have discussed a number of features betraying the ambiguous feeling Shona people have vis-à-vis their ancestral spirits, and suggested that this ambiguity derives from the fact that in the ancestral spirit concept they have welded together two fundamental experiences of their daily lives: the unpredictability of every man's fate,[78] and the power which is inherent in Shona holders of authority and property.

Perhaps the description of the concept up to this point has given the impression that the Shonas' ambiguous attitude towards their ancestral spirits is overwhelmingly characterized by apprehension. In fact, as the term ambiguity implies, and as I have already pointed out, the Shona do expect their ancestral spirits to guard them and to bestow blessings upon them. It will be noticed that in the remainder of the chapter this other side of the ambiguous nature of the ancestral spirit concept is more prominently dealt with. I will first summarize the text of three Shona myths and then show how what I see to be important aspects of their symbolic message, still forms part of contemporary Shona rites of appeasement of ancestral spirits.

Shona myths: Life from death

In attempting to grasp the message or messages contained in myths, one must, of course, be aware that myths are highly complex phenomena. I do not claim to lay bare all of the meanings which may be enshrined in these three myths, neither is there a need to do this. All I want to do, is to point out emphases in them which have striking parallels in contemporary Shona ritual, which is performed to 'settle' the ancestral spirits in the spirit world. It would seem that the similarity between notions expressed in these myths and in the rituals is too great to be merely fortuitous.

1. Mwari made Mwedzi (Moon) at the bottom of a pool (*dziva*) and gave him a medicine-horn (*ngona*). In spite of Mwari's warning that he would be sorry for it, Mwedzi wanted to go on to the dry land which was barren and lifeless. Upon Mwedzi's bitter complaints, Mwari gave him a girl, Massassi (Morning Star), who was to be his wife for two years. The girl was given fire-making tools. The two spent the night in a cave, where they kindled a fire and lay down on either side of it. Mwedzi dipped his finger into the medicine-horn and touched Massassi's body with its contents. As a result, Massassi gave birth to grass, bushes, and trees. The trees grew until they reached the sky, whereupon it started to rain. Mwedzi and Massassi now lived in plenty, but after the stipulated time was over, Massassi had to return to the pool. Mwari however, gave Morongo (Evening Star) to Mwedzi, who again was to be his wife for two years. Each night the two anointed themselves with the contents of Mwedzi's medicine-horn and in the morning Morongo gave birth to all kinds of animals. Eventually she bore also boys and girls. These had grown up when evening came. When Mwedzi wanted to continue to sleep with Morongo she said to him: "Look, your daughters are grown up. Sleep with them". He did as she had told him and he became king (*Mambo*) of a great people.

   Morongo now slept with the snake who had its lair under her bed and she no longer bore children. One day Mwedzi came to sleep with Morongo against her will and was bitten by the snake. He fell ill and the next day it did not rain. A terrible drought ensued and Mwedzi's children consulted their divining bones in order to find out what they should do. They were told to send the ailing king back into the pool. They strangled Mwedzi and buried him together with Morongo. Then they chose another man to be their king.[79]
2. Once it had not rained for a whole year. The diviner-healer said that one of the king's daughers who had never slept with a man was to be sacrificed. However, it was found that none of the king's daughters of marriageable age was still a virgin. Then the diviner-healer ordered that a small girl of the king's household was to be kept within an enclosure (*rushanga*) until she would be fully grown. The enclosure was built of stones around a huge anthill on top of which there stood a big tree. The girl's grown-up sisters brought her food every day and kept watch that no man could come near her.

   It took two years before the girl had become a mature maiden. During all this time there had been no rain and many people had died of starvation. Now the people were summoned to the enclosure and the diviner-healer opened its entrance. A hole was dug in the anthill, between the roots of the big tree. The people called on the ancestral spirits and strangled the girl. While they danced around the enclosure, the diviner-healer buried her in the hole which had been dug in the anthill. Immediately the tree started to grow. It kept growing as long as the people were dancing. Eventually it reached the sky and a strong wind began to blow. The leaves of the tree turned into clouds and it rained for thirty days.[80]

3. A man had a barren wife. One day he caught a fish which pleaded with him to let him return into the water. The man agreed, after the fish had told him how he could cure his wife's barrenness. He was to bury two seeds of the *muhacha* tree one after the other in the ground. Each time he did so the seeds sprouted and his wife became pregnant. She bore him two sons.
After their parents had died, the older of the two sons went to live at the king's village. When he arrived there the people were busy with the rain-making ceremony as there had been a drought. A beautiful daughter of the king was locked up in an enclosure which had an opening towards a lake. Then everybody went home but the young man stayed behind. He asked the king's daughter to let him in, but she replied that she was to be fetched by the fish with six heads at midnight. After this it would rain again.
At midnight the young man climbed over the enclosure and when the fish emerged from the water he cut off its six heads. Immediately it started to rain. As the rain began to fall, the people rejoiced, saying: "The fish with six heads has eaten the king's daughter, but it has given us rain". When they arrived at the enclosure they found the dead fish and the two young people. They brought them to the king who upon hearing what had happened gave his daughter to the young man as his wife.[81]

A striking characteristic of these myths is their apparently optimistic outlook on life. It is the archetypal ancestors, Mwedzi and his two successive female associates, who procreate all life-sustaining substance covering the earth. The emphasis is constantly on growth. It finds expression in the recurrent pregnancies of Massassi and Morongo, in the tree's growing into the sky from where it 'fetches' the rain, and in the miraculous pace at which Mwedzi's children grow into a people. One should here recall the intimate connection which the word *midzimu* has with the notion of having the 'power of growing up', by the fact that it belongs to the third noun class.[82]

Death decidedly has a place in these myths. However, man, while being afraid of death, is bent on making the best of his life-span, even to the extent of going against Mwari's injunctions. When the Mambo finally meets his violent death at the hands of his own descendants, even this deed is interpreted as his being sent back into the pool from whence he had originated. His being called 'Moon' is also a pointer that Mwedzi, the author of life, never really dies. In the same way in which the moon 'dies' and 'rises' in constant succession, life will go on, although no generation of men can escape the dreadful experience of drought, during which death's harvest is often particularly plentiful, and even the strongest ultimately become death's victims. Although the first myth does not positively say that Mwedzi's strangulation broke the drought which had been caused by his illness, this conclusion is implied. In the two other myths, however, the life-giving death of a human victim is the main theme. One of the king's daughters, whose potential fertility is preserved intact, must be sacrificed to end the drought and to make it rain. The place where the sacrifice is to take place is surrounded by an enclosure which will again be encountered in my discussion of the 'bringing-home' ritual.

In both the second and third myths, the enclosure, apart from the fact that it enshrines the virgin-victim, is linked with other symbols of fertility. In the second myth it is built around an anthill with a large tree, whose powers to grow increase enormously as a result of the victim's death.[83] In the third myth the enclosure borders on a lake which recalls the pool which has been encountered in the first myth and earlier in this chapter. Even in the third myth the theme of the tree with prodigious powers of fertility is not missing, although the tree is not positioned inside the enclosure. However, the third myth links the birth of its hero, who is responsible for causing it to rain, directly with the planting of a *muhacha* tree. It will soon be seen that the 'bringing-home' ritual also contains a striking parallel to the burial, in the second myth, of the strangled girl between the roots of the tall tree. One ought also to keep in mind the effect this burial had on the mythological tree.

The mutuwa tree

It is to be expected that a number of trees in the bushveld have related qualities and could, therefore, take the place of the *muhacha* tree (*Parinari curatellifolia*) in the third myth. It will be noticed that in the second myth no name is given to the tree in the centre of the enclosure. Naturally, in recounting the myth, people in each area would mention the tree which is most common in their part of the country. My description of the 'bringing-home' ritual is based on Korekore ethnography and the tree figuring in these ceremonies is the *mutuwa* (*Kirkia acuminata*). Both botanically and visually this tree and the *muhacha* tree are closely related to each other.[84]

I noted in chapter two that in certain areas of Shonaland the *mutuwa* tree figures prominently during rain ceremonies which are held in honour of the territorial spirits. The sacred grove of the famous rain spirit, Karuva, situated to the north-east of the Korekore, was, until recently, a major ritual focus for them. The grove is remarkable because of the large number of *mutuwa* trees which grow around its central shrine.[85] Bourdillon writes that in spite of its alternative name, *mutimuchena* ('the white tree'), the *mutuwa* is not particularly striking on account of its colour. Its pale grey bark is barely 'whiter' than that of most of the other trees growing around it. He suggests that the height of the *mutuwa* – (it stands out taller and fuller than the normal, rather stunted, woodland vegetation) – may have given the tree the ritual significance it shares with the giant baobabs, which are also used for spirit shrines. 'Or perhaps', he says, 'it obtains its significance from the trees at the sacred grove of Karuva' All his attempts at obtaining reasons for the ritual significance of the *mutuwa* tree from informants, failed. 'The only reason they could give for going to the *mutuwa* tree rather than to any other was that they were told to do so by their seniors who are all dead now'.[86]

Bourdillon says further that 'there are no well-known myths about the world beyond or before the chiefdom and its related neighbours'.[87] Frobenius collected the above myths forty years before Bourdillon did his research among Korekore of the Diwa and Chiruja chiefdoms. Thus he had the advantage that he could still encounter elders whose youth had been in precolonial times. However, he too complained about the lack of interest in 'transcendence' among some of his informants.[88] In several places in his book he hints at the possibility that people might have dissimulated before him their knowledge of essential Shona mythology for fear of reprisals from White administrators because of the reference in some myths, to human sacrifice.[89]

There is no reason to doubt the professed ignorance of Bourdillon's informants as regards the deep significance of the rituals which they perform for the appeasement of their ancestral spirits.[90] Nevertheless, as has been seen, Korekore and other Shona peoples still uphold the 'hot' and 'cool' concepts and connect them with a variety of situations involving their ancestral spirits. As in other African societies,[91] for the Shona too, the 'hotness' of wrath and anger is coupled with the colour red, and the 'coolness' of peace and calmness with the colour white.

The *mutuwa* tree is expressly linked with this red/white, i.e. 'hot' 'cool', colour classification. The final rite for 'bringing home' the ancestral spirit of a chief, and any other rite addressed to the ancestral spirits, may not take place during early spring when leaves still have a reddish-brown colouring.[92] Apart from this express statement, there are other relevant observations to be made with regard to the *mutuwa* tree, which cannot escape the present-day Shona either.[93] As I have noted, the tree is for them 'the white tree', although the bark has a more greyish colour. It is very likely, therefore, that the white referred to concerns the comparatively wide ring of white sapwood around the *mutuwa's* dark, coppery brown heartwood with its attractive, wavy markings. It may also be significant that this 'red-hearted' tree bears long stalks with loose bunches of insignificant *white* flowers, which appear together with the *red* leaves. Another of its characteristics is that it can be grown easily from a seed, that it grows very fast, and propagates readily from pole-cuttings.[94]

Let us now consider the rituals of re-integration and appeasement which the north-eastern Korekore perform for their ancestral spirits.

Appeasement and 'bringing-home' rituals

There is, first of all, the *kokumusoro* or *mutotwe* ritual, which is performed on the evening following the burial.[95] Reserving discussion of its first part for later, I turn immediately to its second part, which follows upon a night of dancing in the homestead of the newly buried. On the following morning, a 'daughter-in-law' of the deceased places on her head a pot of thin porridge (which had figured prominently during the first part of the ritual), goes on her knees, and moves slowly from the homestead to the new grave. She is followed by another 'daughter-in-law' carrying a pot of water, and by the rest of the mourners. While the procession in Diwa moves in the described manner to the grave of the newly deceased, in Chiruja it proceeds instead to a *mutuwa* tree in the bush, near the homestead of the deceased. The size of the tree is unimportant, but it must not have been used previously for ritual purposes. After the procession has reached the tree, a 'wife-receiver' (*mukuwasha*) clears the base of the *mutuwa* and the two pots are placed to one side of it. All sit down and salute the temporary shrine by ritually clapping their hands, while the 'daughters-in-law' lie down feigning sleep.

The master of ceremonies (*mutonje*) indicates a spot at the base of the tree in which the deceased's 'ritual friend' (*sahira*) has to dig a hole. When he has finished digging the hole, the 'ritual friend' draws a small quantity from the thin porridge with a drinking gourd, takes it to his lips, and then pours it into the hole, saying: 'I am your *sahira*. I have mourned you today. Do not forget me'. Then he claps his hands in salutation. After him, all those present do likewise. A few make longer addresses, asking the ancestors to welcome the new ancestral spirit, and the latter to look after his children and relatives. The water in the second pot is used by the people to make a ritual ablution after each of them has made his address.

Secondly, there is the *bona* ritual,[96] which the north-eastern Korekore perform one to two years after the burial of the deceased. Its intention is to terminate the ancestral spirit's free roaming about and to settle it permanently as a 'guardian' in its former homestead. Significantly, however, in most instances the ceremony is only performed at the spirit's specific request. Again, I concentrate firstly on those aspects of the ritual whose close relationship to essential points of the above myths is obvious.

On the eve of the ceremony, the 'wife-receivers' (*vakuwasha*) have to provide the senior agnatic member of the deceased's family with a pure black male sheep or goat, referred to as *ngombe*,[97] which must have at least begun to grow its horns. This animal is then consecrated by the deceased's 'ritual friend', who pours a cup of fermenting beer on its back and formally addresses the ancestral spirit: 'N., we bring you home today'. After this dedication ceremony the animal is called by the name of the deceased.[98] One of the 'wife-receivers' ties a small strip of new white cloth around its neck which must remain there until after its death.

The next morning, some male and female 'wife-receivers' carry on their knees the dedicated animal, as well as pots of beer and water, to the edge of the homestead clearing. Other 'daughters-in-law' sweep the way before them, using their bare hands. From there everybody proceeds on foot to a selected *mutuwa* tree in the bush, joining in the singing of mourning chants. Having arrived at the tree, the master of ceremonies marks a spot at the foot of the tree with the freshly cut *mutuwa* stake which had been used during the night to tether the dedicated animal in the middle of the homestead. Using this stake, the deceased's 'ritual friend' digs a hole, fills it with water and millet beer, and, helped by a 'wife-receiver', holds the animal's head into the puddle until it is drowned. Just before the animal is drowned, there is a moment of tension among the participants in the ritual, for if the victim should bleat this would be taken as a sign that the ancestral spirit refuses to come home.
After the dedicated animal has died, the 'ritual friend' covers it completely with a new white cloth and the master of ceremonies places on its head a token gift in the form of a small coin. The 'wife-receivers' place the dead animal on a leafy branch and cut small pieces of flesh from all parts of the carcass. These are first roasted over a small fire, then cut even smaller, and handed to the master of ceremonies. Everybody gathers now around the hole which in the meantime has been surrounded with a rough enclosure (*rushanga*). Starting with the master of ceremonies, each person takes a piece of the roasted meat to his mouth to 'taste' it, and then drops it into the hole, at the same time addressing the spirit of the deceased. Afterwards, waste matter from the slaughtered animal and any objects which have had contact with the victim are deposited in the enclosure. Then the hole is closed and all salute the shrine.

> Now comes the most important part of the ceremony and an atmosphere of anxiety and tension becomes apparent. A 'son-in-law' must cut a small branch from the *mutuwa* tree with one, clean stroke of his axe. Should he fail, it means that the spirit refuses to co-operate; it has a grudge against this 'son-in-law' and another has to try. When he has cut the branch, he must immediately cover the cut end with his hand to prevent the spirit from coming out. The *sahira* takes the cloth which had been used to cover the dead animal and, when he is ready, the 'son-in-law' removes his hand to let the *sahira* hurriedly cover the cut end with the cloth. The 'son-in-law' holds the cloth over the end and

> holds the branch on the ground. Another 'son-in-law' places the skin of the beast over the leaves of the branch and the procession back to the village begins.[99]

During its homeward-bound journey, the ancestral spirit is frequently 'cooled' by a 'wife-receiver' who sprinkles beer over the branch. The master of ceremonies sees to it that the tether, tethering stake, the strip of white cloth from around the animal's neck, its horns, and some of its stomach contents are brought back to the homestead. Having arrived there, the procession receives an enthusiastic welcome from the young people who had to stay behind. While the 'daughters-in-law' lie down in reverence, the *mutuwa* sapling is dragged through the doorway of the hut belonging to the senior member of the family. It is immediately pulled out again. The 'ritual friend' digs a small hole by the doorway, using the same stake which he had employed for the purpose when he dug the hole in the bush. He drops the slaughtered animal's stomach waste into the hole, plants the cut end of the branch in it, and fills the hole quickly, as soon as the cloth is removed. The branch is fastened to the wall of the hut, together with the ceremonial articles which have been brought back from the bush. After this the 'daughters-in-law' are allowed to stand up. The whole assembly salute the ancestral spirit, clapping their hands, and this is often followed by a chant before the branch. Thus the *bona* ceremony ends.

Analysis of symbolism involved in the rituals

It is quite obvious that there exist a number of cognitive parallels between the two rituals which I have just described and the second and third myths recounted above.

The rough enclosure made during the rituals, surrounds a spot, as does the enclosure figuring in the myths, where mortal man endeavours to prevail over the laws of nature, including that law which decrees the inevitability of his own death. Droughts and their accompanying misfortunes were broken in mythological times through the 'sending back' of the first *Mambo* into the pool. The sacrificial death of the king's daughter in the second myth, and her burial between the roots of the tree inside the enclosure, is the most striking parallel to the rituals of re-integration which the Korekore perform for their ancestral spirits at the foot of a *mutuwa* tree. In the third myth the enclosure is adjacent to a lake and the king's daughter is expected to meet her death through being killed by a mythological creature emerging from this 'pool'. In each case the result of these sacrificial deaths is the return to normality: abnormal dryness and heat, and the misery they cause, are followed by an abundance of rain and fertile growth.

The performance of the *kokumusoro* ritual in Diwa at the grave of the newly buried, rather than at the base of a *mutuwa* tree, emphasizes the conceptual identification which is made between the mythological figure and the dead ancestor-to-be of today. This identification is dramatically underlined by the fact that during the 'bringing-home' ritual the ancestral spirit, who is conceptually presumed to be present in the dedicated animal,[100] breathes his last into the 'pool' which has been created in the hole for this purpose. The same symbolic identification between the mythological archetype and today's deceased, or rather his ancestral spirit, is expressed in several other actions taking place during the rituals, like the 'burial' of libations dedicated to the ancestral spirit and of parts of the dedicated animal, in the 'pool', and, especially, the symbolic return of the deceased to his homestead inside the mutuwa branch and his ritual 'burial' there at the entrance to the senior's hut.

It is reasonable to assume that the Shona, on the conceptual level, link the common rites which I have described to significant ceremonies of the past and present involving their chiefs. This is not at all surprising if one remembers the close symbolic connection which exists between the chief, the people, and the land, as was seen in chapter two. Firstly, Eastern Korekore make the revealing statement that their *bona* ritual is equivalent to the chiefly succession ceremony which takes place at the grave of the last reigning chief.[101] During it, the newly elected chief has to sit on his deceased predecessor's grave and a sheep is drowned in a ritual pool of beer, in the same way in which the dedicated animal is killed during the 'bringing home' ceremony. Secondly, it has been seen that the leaves of the *mutuwa* branch which is dragged home from the bush are covered with the skin of the drowned animal victim. This is reminiscent of the practice of burying the mummified body of a dead chief wrapped in the skin of a black bull. When the chief's body is carried to the cave in which it will find its last rest, it is set down on the ground at regular intervals and saluted.[102] Even this custom of interrupting the dead chief's journey to his final resting place through a number of stops has its counterpart in the *kokumusoro* and 'bringing home' rituals of the Korekore: during the processions to and from the grave or the enclosure, the group periodically halts and salutes the deceased (who is symbolically represented in the ritual objects) by placing token payments on the heads of the 'wife-receivers' who carry them. Also during these halts 'daughters-in-law' have to lie down in the previously described manner.[103] The full implication of the Korekore's saying that their 'bringing home' ritual is equivalent to the succession ritual of chiefs, becomes clear when one takes a look at what follows that ritual. During the inheritance ceremony, which ideally should follow immediately upon the 'bringing home' ritual, the deceased is supposed to be resurrected in the person who inherits his possessions. In this ceremony, the inheritor and the deceased's widow(s) sit together on a mat, sipping beer. The women present shrill in salutation. Then the master of ceremonies, speaking in the loud tone of voice used to address spirits, says to the inheritor:

> "N . . . , you have arisen today", and goes on to charge him with all the family responsibilities of the deceased, exhorting him to perform his duties properly. He then claps his hands in respect while the female *sahira* [ritual friend] shrills. All the attendants present token gifts . . . for themselves and for absent relatives, addressing the inheritor with the same formula, but more briefly. Finally, the 'ritual friend' does the same with a long exhortation to the inheritor.[104]

After this the possessions of the deceased are laid out before him and sprinkled with 'medicated' water.

Lastly, a closer look has to be taken at the other important focus of ritual attention during the *kokumusoro* and 'bringing home' rituals, namely a pot of thin porridge or fermenting beer. Like the animal victim which is drowned in the 'pool', the pot is identified with the ancestral spirit and is thus conceptually 'hot'. For this reason, it is cognitively classified with things that are red, whatever its true colour may be. On the evening before the pot is carried in procession to the grave or to the enclosure in the bush, it is at the centre of the ritual which is performed in the deceased's homestead. 'Cooling' *depa*[105] saplings are planted into the ground to form a circle around the pot. Then the pot is sprinkled with meal, whose whiteness has been seen to be cognitively linked with the state of peace, calm, and coolness.[106]

All through the night the people dance around this temporary 'enclosure'. The following morning, the pot is smeared with a mixture of meal and water. Afterwards, it is respectfully washed by the 'daughter-in-law' who subsequently carries it to the next ceremonial scene. The same ritual sprinkling and smearing of the 'red' pot with white meal takes place during any further ritual of appeasement which may be requested by the ancestral spirit before or after the 'bringing home' ritual has been observed for him.[107]

The colour symbolism which is involved in the sprinkling and smearing of the pot with meal gives a hint as to why the *mutuwa* tree should hold such an important place in the Shona's rites of re-integration and appeasement for their ancestral spirits. The texture of *mutuwa* wood can be seen as an image of the aim of the 'bringing home' and appeasement ceremonies. That is a potentially dangerous 'hot' spirit, (represented by the red core of the *mutuwa*), is 'cooled' and cajoled into returning to the restful peace of his former abode among the living (represented by the large ring of white sapwood around the *mutuwa's* red core). The same symbolism is also contained in the white strip of cloth being fastened around the neck of the dedicated animal, as well as in the covering with a white cloth of the dead victim and of the severed *mutuwa* branch which is dragged to the homestead.

The intrinsic meaning which the colours red and the white have for the Korekore expresses symbolically the sense of ambiguity which they experience vis-à-vis their recently departed, and particularly during the 'bringing home' ritual, for

> on the one hand, the *bona* ceremony is held in honour of the spirit and is frequently instigated at the supposed request of the spirit. Yet the people are constantly afraid that the spirit will not co-operate.[108]

However, one should not overlook the fact that the symbolism is located in a tree noteworthy for its extraordinary powers of growth and propagation. Thus, even if present-day Korekore should no longer have – nor indeed if they should never have had – any knowledge of the myths I have quoted, the rituals alone, and the symbolism implied therein, would still seem to contain a message for them.[109] If one were to try to put into words what this message might be, it could be something like this: 'Although death has robbed us,[110] our deceased member not only lives on, but he cares for us and helps to maintain our society's time-honoured structure of authority and ownership.' That this last aspect is of prime importance for understanding the ambiguous nature of the ancestral spirit concept I have already shown in the first part of this chapter. It is also apparent from the subservient roles which the 'wife-receivers' have to fulfill throughout the funerary rituals.

NOTES

1. Singular: *mudzimu*.
2. Hannan (1974), p. 372.
3. Plural of 1st class nouns.
4. Hannan (1974), p. 363. Cf. also *ibid.*, p. 372: *Mudzi* [Proto-Bantu], (class 3) = Root, herbal medicine.
5. Murphree (1965), p. 103. He bases himself on Torrend (1921). Bourdillon (1972a) adopts Hannan's classification of *mudzimu* as a word belonging to class 1, but then speaks of *midzimu* throughout his thesis. He says expressly (p. VIII) that *zvemidzimu* means 'things concerning the spirits'. Murphree, *loc. cit.*, suggests that '*midzimu* is the general term, indicating a collective unity, while *vadzimu* is more personal and is used by the individual to refer particularly to those spirits of his patrilineage who are remembered by name'.
6. Bourdillon (1968), p. 108-50: *passim*.
7. Beattie (1967), p. 256. Brain, whose argument against Kopytoff over the use of the expression 'spirit-elders' I am going to present shortly, says that it may seem unnecessarily pedantic to point out that numerous ethnographers have inadvertently used the personal noun class prefix in connection with the respective word for 'spirits of the dead'. The important point is, however, that 'if people do, as Kopytoff tells us, really regard ancestors as not being qualitatively different from elders, why then assign the name for them to a non-person noun class?' (Brain [1973], p. 123).
8. *Mubvuri* (3rd class). 'When the shadow of a person has gone he has become a spirit': Hannan (1974), p. 363.
9. The question receives more detailed attention in another subsection of this chapter. Cf. below, p. 64.
10. According to Magava (1973, p. 152) children who have reached the teething stage are regarded as grown-ups, however, the 'bringing-home' ceremony is only performed for persons who have procreated children: cf. Bourdillon (1968), p. 50.

11. The meaning of this expression and the ceremony it involves will be discussed below in this chapter.
12. Bourdillon (1976), p. 23.
13. Cf. Holleman (1953), p. 2; Bourdillon (1972a), p. 218.
14. Bourdillon (1976), pp. 237, 246.
15. Bourdillon (1972a), p. 192.
16. Cf. Hammond-Tooke (1974), p. 320; Baumann (1936), p. 219f.
17. Cf. Gelfand (1959), pp. 14, 30-2; Daneel (1970), *passim.*
18. Frobenius [1930], p. 149-59 & *passim;* Gelfand (1964), p. 62f. As *Dzivaguru*, 'The Great Pool', Mwari is conceptually linked with the people's expectation of rain: cf. Daneel (1971), p. 81. Cf. also Heintze (1967), and the myths recorded below in this chapter.
19. The meaning of this term will be discussed in a special subsection of this chapter.
20. Cf. Bourdillon (1976), p. 237f; also Beidelman (1973), p. 142.
21. In the last subsection of this chapter.
22. Bourdillon (1972a), p. 210.
23. *Ibid.*, pp. 181, 188f, 190; Bere-Chikara (1970), p. 20. The mourners' washing themselves may well primarily serve the purpose of removing the imagined contaminating effect of death, but the action would also seem to point to the heat syndrome being involved in the situation. Cf. also Magava (1973), p. 155.
24. Bourdillon (1972a), pp. 192, 210, 230, 234.
25. Alias 'to burn the spirit': *ibid.*, p. 216.
26. *Ibid.*, p. 192. Prof. J. Goody has attempted to demonstrate the intricate links existing between death, property, and the ancestors, in his book of that title. Although he has been rightly criticized for his failure in the book 'to bear out the observer's point of view [merely] by the actors' statements' (H. Keesing [1970], p. 41), his interpretation of his ethnographic material, using 'psycho-physical factors which are not contained in the actors' interpretation', are nevertheless highly perspicacious. Thus he writes: 'the major contradiction here lies in dismissing the dead with honour and yet retaining their property without loss . . . The offering of token amounts is one partial solution; the concept of a joint corporation such as a lineage, consisting of both the living and the dead, is another. But neither of these ideas altogether solves the central difficulty that is reflected in the fear which the living have that the dead will attempt to perpetuate their network of social ties in ways dangerous to the survivors'. Goody (1962), p. 73f.
27. Magava (1973), p. 154f. For identical reasons, the Lodagaa of West Africa surround a deceased with all his possessions while he is exposed on the funeral stand: Goody (1962), p 203. Concerning the notion that possessions are a part of a person's extended social personality, cf. *ibid.*, p. 200, and Evans-Pritchard (1973), p. 94.
28. There is the occasional exception to this rule, particularly if the deceased has left behind very young widows who might be tempted to 'burn his grave'. Cf Bourdillon (1972a), p. 216; (1976), p. 249.
29. Bourdillon (1976), p. 249.

30. Bourdillon (1972a), p. 239.
31. Kopytoff (1971), p. 140.
32. Brain (1973), p. 124f. Bourdillon found that the Eastern Korekore, among whom he did his fieldwork, know of an impersonal 'frightening power to kill which certain persons may have or acquire', referred to as *nzimu* (class 9). (Personal correspondence).
33. *Ibid.*, p. 125.
34. Bourdillon (1968), p. 7. Also what follows relies on *ibid.*, pp. 6-13.
35. Holleman (1952), p. 66. Holleman's spelling of Shona words does not always agree with the present orthography. I have rendered Shona words quoted by him in his spelling.
36. Holleman (1952), pp. 67, 389.
37. Bourdillon (1968), p. 55, basing himself on Garbett (1960), p. 14.
38. Bourdillon (1972a), p. 10 Later in this chapter it will be seen that the same adage is relevant to the ritual function of cattle.
39. Aquina (1973), p. 53. Numbers inserted by myself.
40. Garbett (1963), p. 106.
41. Bourdillon (1972a), p. 356. Cf. also Bourdillon (1976), p. 256; Crawford (1967), p. 80.
42. Bourdillon (1976), p. 47.
43. Bourdillon (1976), p. 47.
44. Aquina (1973), p. 53. Ranger (1976), p. 22, seizes on a more direct statement in the same vein by G. P. Murdock (1959), p. 377, according to which the Shona 'followed only the matrilineal rule of descent in the not very distant past'.
45. Holleman (1952), pp. 63, 67, 352.
46. Prof. Edmund Leach, in his book entitled *Rethinking Anthropology* (1966 edition), p. 4f, has very aptly made the point that the 'type-finding' social anthropologist is often so preoccupied with fitting the evidence encountered in the field into his preconceived models, such as matrilineality, that he is unable to perceive other, and more plausible explanations for the 'problems' created by his models.
47. Cf. Holleman (1952), p. 63 and the diagram on p. 354.
48. Cf. *ibid.*, pp. 28 & 65.
49. Cf. *ibid.*, pp. 218, 242, 255. Bourdillon (1976, p. 254ff.) summarizes the Shona ideas about the 'spiritual aspect of man', stressing that these are fraught with ambiguities and contradictions. However, so much is clear that the 'personal spirit' (*mudzimu*) of a child, whether it is a boy or a girl, is always believed to come from his agnatic line, entering his mother's womb when his life begins there as a snake or a worm.
50. Cf. Also Aquina (1968), p. 50f; Holleman (1952), p. 240.
51. Holleman (1952), pp. 28, 210, 212, 254.
52. Holleman (1952), pp. 213, 321, 353.
53. *Ibid.*, pp. 240, 355.
54. *Ibid.*, p. 351.
55. Holleman, (1952), p. 182.
56. *Ibid.*, p. 70.

57. *Ibid.*, pp. 90, 178.
58. Cf. Daneel (1971), pp. 48, and 161. He also notes (cf. *loc. cit.* and p. 99) that the husband's sister, as *samukadzi* and *vamwene*, 'owns' his wife and has similar 'mystical power' to that of his mother-in-law. Her dissatisfied spirit is able to 'grip the wombs' of her brother's daughters or brother's son's daughters, making them temporarily or permanently barren.
59. Holleman (1952), pp. 182, 351.
60. *Ibid.*, p. 182f. The reason is that women at times dispose of their cattle to buy food for their children, as has been mentioned; that the 'cow of motherhood' may die without leaving offspring; or, in those cases where it is only given after the death of the wife's mother, it may be immediately sacrificed to the latter's ancestral spirit.
61. *Ibid.*, p. 356.
62. *Ibid.*, p. 358.
63. *Ibid.*, p. 357.
64. Holleman (1952), p. 67.
65. *Ibid.*, pp. 278ff.
66. Cf. Bourdillon (1976), p. 48.
67. Cf. Aquina (1973), p. 53, where she quotes a Karanga man as having said to her: 'When the spirits are far away, they cause us no sickness, and there is nothing we have to ask them for. At such times we can almost do what we like. Only when a spirit hovers nearby will he watch us, and punish us when we do wrong. If we suffer misfortunes, we therefore ask the diviner to find out which spirit is near and what he wants from us. When we have given him what he desires, all will be well again'.
68. Cf. Bourdillon (1976), p. 240.
69. Gelfand (1959), p. 153.
70. Bourdillon (1976), p. 273.
71. *Ibid.*, p. 272. For a typical case, cf. Fortune (1975).
72. Bourdillon (1972a), p. 267f.
73. Gelfand (1959), pp. 153-61.
74. Cf. Gelfand (1962), p. 69f. For the purpose, a diviner-healer or kinsman performs a ceremony over the grave of the wronged person.
75. Garbett (1969), p. 123f.
76. Gelfand (1959), p. 153f. It is not clear from his account whether the avenging spirit himself - through a medium - devised the manner in which the son was to be humiliated or whether it was prescribed by the diviner-healer.
77. *Ibid.*, pp. 154-60.
78. It will be seen in chapters four and five, that misfortune of various kinds is by no means solely attributed to the action of ancestral spirits. However, the other agents of 'fate' are intricately linked with these: the stranger spirits which I shall discuss in the next chapter base themselves on the ancestral spirit concept, and I have already mentioned that witchcraft attacks are only believed to be possible if they are permitted by a disgruntled ancestral spirit.
79. Frobenius [1930], p. 237-40.
80. *Ibid.* (1930), p. 204-6.
81. *Ibid.*, p. 179f.

82. Cf. the beginning of this chapter.
83. In one portion of the Mtoko district, where the 'bringing-home' ritual is held at the grave of the deceased, as the party are approaching the grave they are warned to be very quiet, because they are now 'at the anthill': cf. Holleman (1953), p. 14.
84. Cf. Pardy (1951); v. Breitenbach (1965), p. 251f. Murphree (1969), p. 21; 37-9. In Frobenius [1930], p. 197f. the *mbila* rain-making ceremony described in the second myth is clearly associated with the *muhacha* tree, which is prominent in the third myth. Cf. also the plates in Gelfand (1959), opposite pp. 24, 40, 41, and the text on p. 33.
85. Bourdillon (1972a), p. 245.
86. *Ibid.* p. 246.
87. *Ibid.*, p. 355.
88. Frobenius [1930], pp. 132, 136.
89. *Ibid.*, p. 240-244.
90. Cf. Bourdillon (1972a), p. 242. Monica Wilson (1971), p. 72 notes that 'during periods of rapid social change consciousness of symbolism diminishes'.
91. Cf., e.g., Turner (1966), *passim.*
92. Bourdillon (1972a), pp. 99, 245.
93. Cf. Pardy (1952); v. Breitenbach (1965), p. 426-8. Bourdillon himself notes that 'our understanding of their [= the Shona] beliefs must be based on ritual expression and association rather than on verbal theory'. Cf. Bourdillon (1976), p. 258. I am told by Professor J. H. M. Beattie that in Bunyoro the *mutuma* tree has the same qualities as the Rhodesian *mutuwa*, and that it also fulfils a significant role in local rituals.
94. Bourdillon (1972a),p. 246 writes that 'one old man, when pressed on the subject, suggested that they used the *mutuwa* branch [during the 'bringing-home' ceremony] because when it is planted in the ground it "rises" again'. Cf. also the emphasis the three myths quoted above place on fertility.
95. Thus, e.g., in Diwa chiefdom: cf. Bourdillon (1972a), p. 206-11. In neighbouring Chiruja, however, it may be delayed for weeks or even months, although the ideal there is also that it should follow immediately upon the burial: cf. *ibid.*, p. 212-15.
96. Cf. Bourdillon (1972a), p. 220-30.
97. Shona word for 'cattle'. This indicates that in the past, when cattle were more plentiful, Korekore would rather use cattle for this ceremony, as is the general trend elsewhere in Africa. An alternative term for the *bona* ritual is *kutamba ngombe*, 'to play (dance) the beast' or *kutamba mudzimu*: cf. Bourdillon (1972a), p. 220.
98. Also among the Korekore, the heads of most families still dedicate a bull to their father, whose name it is given. The bull is informed of changes in the herd necessitated by the payment of bridewealth etc: cf. *ibid.*, p. 240, also Makamure (1970), p. 16.
99. Bourdillon (1972a), p. 228.
100. The animal's colour, pure black, is commonly in Africa symbolically linked with the dark, heavy clouds promising rain.

101. Cf. Bourdillon (1972a), pp. 94-7; 234.
102. Cf. Frobenius [1930], p. 129-31. Bourdillon tells me this mode of chiefly burial still takes place in some areas of Shonaland. Cf. also Kileff (1970), p. 70.
103. Bourdillon (1972a), pp. 208; 229. I cannot go into the deep meaning and widespread use of token payments among the Shona. However, it should be mentioned in this connection that the shrouding of the dedicated animal and the placing of a token payment on its head during the 'bringing home' ceremonies has exact parallels in Shona burial procedures and that people are aware of these similarities: cf. *ibid.*, p. 234.
104. Bourdillon (1972a), p. 232.
105. A water vine, *Cissus integriflora.*
106. Cf. above p. 58.
107. Cf. Bourdillon (1972a), p. 265.
108. Bourdillon (1972a), p. 239. Cf. also Holleman (1953), p. 32.
109. Having read the manuscript of this book, Bourdillon wrote to me: 'I am a bit wary of putting north-eastern ritual with southern myths as you have done here. But your analysis seems to work all right.' (30 December 1977). Cf. the historical section above, p. 22, concerning the Changamire Mambos, and my own cautious remarks about these myths on pp. 71, 75.
110. '*Mandipamba !*', 'You have robbed me !', is the repeated cry of the Korekore mourners before and during the burial of a person. Cf. Bourdillon (1972a), p. 181.

CHAPTER 4

STRANGER-SPIRITS

In the preceding two chapters I have shown how the Shonas' concepts of territorial and ancestral spirits have enabled them to account for the vagaries of life which they experience on the level of the tribe and family group, and to uphold those values which they regard as basic and impermeable to change. Change, however, has always affected the Shona, and has increasingly done so over the past one hundred years. It is mainly in connection with this fact that one finds them employing the concept of stranger-spirits (*mashavi*).[1] That this concept is not unique to the Shona is apparent from the evidence which has been accumulated in J. Beattie's and J. Middleton's *Spirit Mediumship and Society in Africa.*[2]

As the term with which I have paraphrased *mashavi* suggests, these are spirits of deceased people, unknown to the community in which they have chosen a host. They can also be the spirits of animals. The notion of a spirit being able to possess a person has already been encountered in this book. There is, therefore, no need to present first of all the details of an alleged *shavi* possession. Rather, its peculiar aspects will be pointed out as my analysis of the phenomenon unfolds.

It appears that from their early days Shona have accredited stranger-spirits with the power to bring luck to hunters. Thus Bullock writes that 'a *shavi* can alight on a hunting dog or even a gun and confer success while it abides there'.[3] It is not difficult to see why the stranger-spirit concept should have arisen in connection with the hunt: its result is always fortuitous and some men prove to be better at it than others. Also, it forces a man to penetrate into that realm which he typically associates with the unsettled and unknown powers supposedly roaming in the bush. Thus the Hambukushu of Ngamiland, who have lived for a long time on the periphery of Shona territory, believe in 'ghosts of deceased hunters . . . which appear in the form of animals which particular hunters were famous for hunting.'[4] The belief in the existence of a baboon *shavi* throughout Shona country is another pointer to the fact that the concept of stranger-spirits arises from man's experience of conceptual and physical frontiers across which he trespasses or which are trespassed by alien powers. During a seance, people who are believed to be possessed by a stranger-spirit act as baboons are supposed to behave. They lap up liquids in baboon fashion, crouching on the ground; they climb trees, eat ground-nuts and raw food, and generally mimic baboons to the amusement of the onlookers.[5] One early Shona writer thought that to understand the baboon *shavi*, 'it is necessary to think of the mythological belief that baboons were *banu* (people)', living in caves, who were too lazy to work and to plough. For this reason, in the course of time they deteriorated and became what they are now.[6] Thus there may also have persisted a feeling among the Shona that 'baboons represent a confusion between the categories of humans and animals, both in terms of their appearance and their behaviour'.[7] Baboons often raid people's fields and gardens, causing much damage to them. Men, on the other hand, attempt to poison the baboons, or hunt them with dogs and guns. Therefore, during a seance for baboon *mashavi*, the possessed hosts often savagely beat any dogs they may come across and constantly exhort people not to hunt 'their relatives' in the veld.[8]

Harold von Sicard, a prodigious though unreliable writer on Shona history, has made an attempt to trace the origin of the stranger-spirits to the Lemba, a group of people originating from Sena in Mozambique, famous for their craftmanship as smiths. They are still known as traders to people living as far south as the Venda in the northern Transvaal. Since their clan emblem was *Shave* (Eland), the Portuguese used to refer to them as *vha-shavi*, and *shava* became the Shona verb for trading and selling.[9] In this connection, what Mushoma says about stranger-spirits makes interesting reading. He writes that they are

> spirits of foreigners who have died and were buried in foreign soil. The spirits of such people are always where they were buried and people living in the vicinity of their graves are affected. These foreigners in their lifetime caused one to crave for certain things which they possessed . . . These spirits make the craving so strong in one that it results in illness. One even becomes possessed by such spirits so that one is not one's true self. Then it is necessary to have a *shavi* dance in order to pacify the spirit.[10]

However, the Shona generally explain the occurrence of *shavi* possession among themselves by saying that *shavi* are spirits of strangers who died away from home, or of young unmarried persons. In both cases, the spirits of these people would not have been 'brought home', and therefore roam around, seeking to express themselves through a medium.

Although there exist several different lists of stranger-spirits,[11] it is significant that common to all of them are the *madzviti shavi*, who represent the spirits of slain Amandebele warriors. The idea that 'powers' arise from foreigners whose blood has been spilt on the soil of the country they invaded is indeed very widespread in Africa.[12] Although the Amandebele have not been the only invaders Shona-land has seen in the course of its known history, the memory of their raiding parties is still fresh in the people's minds. However, in each Shona group spirits from other neighbouring peoples also figure as stranger-spirits, and, as in Bunyoro, there are various spirits of 'Europeanness' who make their hosts act in stylized roles, such as a district commissioner lording it over his assistants or a White house-wife being particular over cleanliness.[13]

Enlargement of scale

The conceptual connection of stranger-spirits with famous hunters or perhaps with itinerant traders, and their definite association with the invasion of the land by foreigners and certain undomesticated animals, make it quite clear that one of the reasons why the Shona have *shavi* possession cults is that through these they attempt to accommodate newly arriving facts of experience within their traditional view of the world. While Shona ethnographers, on the whole, provide one with little information on the relationship between the frequency of *shavi* possession and the degree to which people affected by it had been exposed to foreign influences, Lienhardt noted that 'the Dinka whose countries are nearest to foreign influence are those among whom the powers are regarded as most strongly operating'.[14]

In this regard, Professor E. Colson has collected interesting material among the Tonga, who are neighbours of the Shona in Zambia, and who consciously link their *basangu* mediums with the Shona mediums of territorial spirits.[15] Traditionally, their stranger-spirits (*masabe*)[16] also contained a number of animal spirits and spirits of neighbouring peoples. The advent of colonial administration however, caused the sudden appearance of such stranger-spirits among them as 'the police' and 'the carrier'. These were later joined by the 'train' *sabe*, the 'bicycle' *sabe*, and other newly created *masabe*. There occurred, however, a virtual explosion in *masabe* activity with the arrival in the area of all the modern apparatus which was employed during the construction of the Kariba Dam. Suddenly there appeared 'aeroplane' and 'motorboat' *masabe*, as well as *madance* (from the English word 'dance'), and the 'guitar' *sabe*, and many others. Some Tonga were saying that *masabe* possession had now become 'a contagious disease'.[17]

Both Ranger and Horton have spoken in this connection of an 'enlargement of scale' which, in varying degrees, had befallen peoples all over Africa by the turn of the century, forcing them to deal with an ever-growing variety of alien configurations of power. According to Ranger, the first step in coming to grips with these new powers, seems often to have been 'the creation of a dramatic stereotype, expressing what were held to be the essential qualities of the alien group, and acted out through rituals of spirit possession.'[18] Schoffeleers has found that a possession cult originating from traumatic experiences suffered by the Mang-anja people of Southern Malawi during the closing quarter of the 16th century, exists to this day side by side with the M'bona cult, a cult which has helped people in the more recent past to 'comprehend' and digest striking events and major social upheavals.[19]

Remarking on the similarity between the way the Alur conceptualize the recent invasion of their country by spirits 'as mobile as modern man', and the *shavi* cults of the Shona, Southall says that this is more likely to arise from 'structural-functional likeness rather than common genetic origin'.[20] While there is small reason to doubt the correctness of his assertion, the fact is, nevertheless, that possession cults do spread and that they may do so at great speed. There exists, unfortunately, no reliable information concerning the role which was played by the *shavi* cults in the time immediately following the Shona Rebellion. It seems, however, that the initiative was then largely in the hands of the mediums of territorial spirits (*mhondoro*), as far as the interpretation of the alien powers was concerned. But when Bullock was writing about the Shona in 1927, 'the [*shavi*] cult was becoming more popular than jitter-bugging in America, and considerably more harmful'.[21]

As for the Venda, the southern neighbours of the Karanga, many of whom had resided for a considerable length of time among the latter during the last decade of the 19th century, it is stated that possession was very rare among them until 1914. From this time onwards, it spread to them from Rhodesia and gained popularity with extraordinary rapidity.[22] At about the same time, a cult involving possession by the *amandiki* spirits arrived in Zululand, apparently via the Tsonga, who are neighbours to the Shona-speaking Ndau.[23] It is worth noting that in most of the African countries where spirit possession had been previously rare or unknown, a distinct class of specialists for the treatment of its symptoms arose together with the appearance of the new phenomenon, sometimes against the opposition of the traditional diviner-healers.[24] The absence amongst them of a distinct class of diviner-healer specializing in the doctoring of *shavi* possession seems to be a further indication that the Shona have a long-standing tradition of using spirit possession as a means to incorporate symbolically, new influences into their accustomed scheme of life.[25]

Strategies for redefining situations

Recently Professor I. M. Lewis has devoted a book entitled *Ecstatic Religion* to the anthropological study of spirit possession and shamanism. In it he suggests that peripheral possession cults, which he contrasts with central possession cults, have the following three characteristics: firstly, the cults play no direct part in upholding the moral code of the societies in which they occur; secondly, they originate outside of these societies; thirdly, their favourite victims usually belong to peripheral strata of these societies.[26] From what I have stated so far about the stranger-spirits of the Shona, it is clear that the *shavi* cults certainly play no direct part in upholding the moral code of Shona society, as does the people's belief in the powers of territorial and ancestral spirits. They are also peripheral in that the various spirits featuring in the cults are thought to have originated outside the conceptual or physical frontiers of Shona society, as I have already noted. However, Lewis's third mark of peripherality, i.e. that the 'favourite victims' of possession cults are either women deprived of power in a male-dominated society, or 'downtrodden categories of men who are subject to strong discrimination in rigidly stratified societies', cannot be said to apply in these exclusive terms to the Shona.

Some stranger-spirits are said to announce their arrival in children, though they may only 'come out' later, after their hosts have attained a more mature age.[27] Gelfand mentions at least one *shavi* cult whose membership consists exclusively of men,[28] while there are a number of others in which men would seem to be in the majority.[29] Most participants in the popular baboon *shavi* seances are men, and although they would normally drop out of the cult while they are still quite young, no social stigma is attached to their participation in them.[30] A headman may even invite the members of a *shavi* cult to perform a seance in order to 'detect any mystical danger threatening the hamlet and to drive it away'. Many of the songs which are sung during such a dancing seance express the values of living together amicably and warn of the dangers resulting from gossip and backbiting.[31]

However, once these qualifications have been made, it must be pointed out that a number of traits which Lewis and others typically associate with female participants in possession cults are also to be found in Shonaland.[32] Bourdillon notes that the possession cults provide Shona woman with an opportunity to express, under the protection of the dissociated trance state, inclinations which they would not normally be allowed to manifest in public.[33] In this connection, one should mention that women frequently claim to be possessed by several stranger spirits, some of whom may be antagonistic towards each other. Their successive 'coming out' during a seance provides the audience with much amusement, and the hosts, no doubt, with gratification.[34] Also women often claim to be hosts to male spirits, and mimic male roles during their 'possession'. They brandish axes and knobkerries, dress in male attire, and occasionally take the male part in the *dende*, a dance with lewd gestures normally performed by a male and female partner.[35]

Although the seances of a *shavi* cult are simple, their preparation can cost the host's husband and other kinsmen a substantial sum of money. Apart from procuring the required dress and other paraphernalia needed for the new host's initiation into the cult, they have to hire for each seance, skilled drummers or other musicians. They must also provide a sufficient quantity of beer and food for those who attend the ceremony, numbering perhaps two hundred.[36] The stipulation that the host to a stranger-spirit must have nothing to do with these preparations underlines the fact that membership in the cult enhances a member's social status, but it would not necessarily warrant one's seeing the *primary* function of female possession cults among the Shona as serving 'an oblique aggressive strategy' as it would for Lewis.

Gelfand has emphasized the purely 'social' aspect which many of the *shavi* cults seem to have; the members' main preoccupation during the seances is focussed on enjoying the dance and each other's company. They are like 'clubs where men and women unrelated by family ties [can] meet and share their common interests'.[37] Thus initiation into possession cults among the Shona follows a general pattern which is recurrent under vastly different social and religious connotations in Africa and Afro-America.

Once a diviner-healer has diagnosed a person's affliction as being caused by a stranger-spirit, a seance is arranged for his or her initiation, which is attended by all those men or women in the area who had been previously afflicted by that particular type of *shavi* and who are now its permanent host. The senior host presiding over the ceremony has to be specially honoured and is paid a fee.[38] After the initiation, the new *shavi* host may become possessed at any seance for members of his or her particular cult. There exists also among one Shona group a close parallel to the sending by women of stranger-spirits to others, which Elizabeth Colson has claimed takes place among the Tonga.[39] Thus among the Kalanga a girl would remain under the tutelage of the woman who 'raised' the *shumba* spirit in her, and she would even be bound to marry into her initiator's family. The latter's social standing would, therefore, increase with the number of *shumba* hosts she can make into her wards.[40]

From these examples it is obvious that men and women do use their membership in *shavi* cults to enhance their status, at least temporarily, in the eyes of the public. However, Lewis's assumption that women join the cults mainly because as legal minors they occupy a peripheral position in traditional society, and that male members of the cults are commonly from downtrodden categories, does not apply to the *shavi* cults. As regards membership in *shavi* cults, it has been expressly stated that there are no apparent characteristics common to all members of a particular cult; further, that these comprise women from all ranks; and that although a woman with a particularly uninhibited personality is likely to join, cult members also include retiring personalities.[41] Quite contrary to the general trend in Lewis's model, female cult members are often said to be the wives of high-ranking men, whose status is enhanced through their ritual roles in the cult and by the powerful support they are supposed to derive from their *shavi*.[42]

It would seem that the definition of spirit possession proposed by Roger Gomm and based on fieldwork in Kenya, suits the Shona evidence better than that of Lewis. For Gomm these cults represent 'a set of strategies for redefining situations' which is available to all.[43] They can be used by husbands to manipulate their wives, by fathers to manipulate their daughters, and by wives to manipulate their husbands, although the extent to which such manipulation is feasible in a given situation may be rather limited. Gomm's definition would seem to cover yet another aspect of the *shavi* possession cults, whose principal function, according to Bourdillon, is to accommodate all those 'powers which appear to impinge on their [the Shonas'] experience without fitting readily into the social system of daily life'.[44]

I have mentioned in the beginning of this chapter that stranger-spirits also serve as a conceptual aid to explain special skills which a person may have, for instance, in the pursuit of hunting. Among present-day Shona, far more important are those stranger-spirits who are believed to enable their hosts to become successful diviners and healers. They are treated with great respect and sometimes rituals are performed in their honour which compare them in importance to the territorial spirits.[45] Although it is believed that baboon *shavi* can also convey healing skills on their hosts, divining and healing spirits are said to 'come out' normally only during private consultations, and never to take part in the often rather frivolous public dancing seances.[46]

Among the Ndau it is even presumed that each person hosts a number of secret stranger-spirits who give him his personal talents and whom he honours in private.[47] Even though all Shona do not share this extreme viewpoint, it is obvious that the *shavi* concept can be used both to explain the special gifts possessed by one person, and to explain away the inferiority of another who does not possess them. In a society in which normality is the ideal, special skills are thus accounted for in a plausible and socially approved manner. 'Possessed' by his *shavi*, its owner is considered to be directed by an in-dwelling external force, when in fact it is aspects of his own personality, developed from personal experience, which he brings to bear on a specific situation. In the introduction to this book, I underline the importance of conceptual separation of subject and object in experience, which the Shona, like the Dinka and other peoples in Africa, achieve through their belief in the existence of stranger-spirits and other ultra-human powers.

'Possession' as a stereotype

Having shown the varied uses to which Shona people put their belief in possession by stranger-spirits, I turn now to a discussion of the 'reality' of possession. In doing so, it is important to make a clear distinction between the phenomenon as a concept, and its physical expression in the person of the 'possessed'. I argue that as a concept, *shavi* possession is a construct of the human mind. As has been seen it entails the belief that the 'possessed' temporarily becomes host to an ultra-human power which 'comes out' in him.[48] That which is expressed on the conceptual level through the image of 'possession', finds its dramatically appropriate expression on the physical level through the 'possessed' person's entering into, or at least feigning to enter into, a trance state.

To speak of a state of dissociation, as many writers on the subject have preferred to do,[49] makes the point involved even clearer: the 'possessed' persons are not supposed to be their normal selves; they must visibly move out of the normal order of things and become altogether different persons. For this reason, when the seance is over, they should not have any memory of what has been going on while they were in trance. There is no reason to doubt that many of the 'possessed' do in fact enter into a highly dissociated state during a seance. However, numerous authors have observed that participants in possession rituals often do not actually fall into trance, but merely simulate dissociation in a more or less convincing manner. Thus, for example, Bourdillon writes that 'possessed' female members of *shavi* cults among the Korekore 'remain very much aware of what is going on around them and are quick to notice anything slightly out of the ordinary'.[50] Just as 'possession' is a conceptual stereotype which the Shona use to explain the aetiology of certain powers whose influence they experience in their daily lives, so is 'dissociation' a stereotype which allows the 'possessed' host a certain amount of freedom with regard to the manner in which he portrays this state, as long as he remains within the generally accepted bounds of credibility.

As regards the Shona, others have adequately described[51] the manner of a *shavi* medium's calling, his or her initiation into a *shavi* possession cult, and the way in which a possession seance is conducted, and there is no need to repeat the description of these observable facts. However, in the remainder of this chapter, I will attempt to deepen understanding of the stranger-spirit concept by discussing a few points which do not seem to be part of the arguments which have been presented by previous writers on the *shavi* possession cults.

Firstly, my claim that 'possession' among the Shona is a well-known stereotype, whether its application makes a person into the host of a territorial spirit, a stranger spirit, or an ancestral spirit,[52] finds support in the fact that children freely attend most of the ceremonies which are held in honour of these spirits. They are thus exposed from an early age to the characteristic, throbbing rhythms of these ceremonies – an important factor which I am to consider shortly. Thus they are also able to watch the typical behaviour and actions of the 'possessed'.[53] Besides, where distinction and honour go with a particular type of 'possession', the desire to strive for its attainment in later life will already be implanted in them at this impressive age.[54]

Secondly, I have noted that every traditional Shona knows from experience that it is best for one to reach an accommodation with the powers that be.[55] Superiors may be manipulated and placated, but they should never be opposed. This experience would seem to account for the common saying among the Shona that 'a spirit will always have its way', a saying which is illustrated by accounts about persons who unsuccessfully tried to resist the demands of their spirits.[56] This stands in contrast with some societies in Africa, in which the aim of the 'possessed' is always to get rid of his spirit by having it exorcised, rather than by accommodating himself to it by undergoing an initiation ritual.[57] It should, however, be noted that the host's coming to terms with the invading spirit is the established pattern in many other African societies.[58]

What is conceptually understood as the process of a person's 'coming to terms' with his spirit, has its visible concomitant in the growing skill with which the host manages to slip in and out of the 'possessed' state. For example, at times it proves rather difficult to induce the trance state in a neophyte, and several seances may be required before his spirit 'comes out' in him.[59] Elsewhere in Africa, novices in 'possession' are often said to be seized with far greater violence than experienced veterans.[60] It can be assumed that Shona mediums too take some conditioning before they are able to control the bodily jerks and contortions which are characteristic at the onset of 'possession'. Indeed, with regard to the Shona, it is said that depending on the type of spirit and the experience of the medium, 'possession' may be induced by the attendants merely clapping their hands in rhythmic fashion or simply giving the medium a little snuff tobacco.[61]

Thirdly, as I have suggested above, there is the prolonged infectious rhythm of vigorous bodily movements; a particularly suggestive type of drumming, which is accompanied by rhythmic singing and hand-clapping, and often underscored through the use of leg rattles. All this produces a cumulative power of suggestion which works both on the *shavi* hosts and on the audience. During a seance, the latter often join the 'possessed' hosts in their dances and occasionally an uninitiated spectator falls spontaneously into a trance.[62] At a seance, the initiate should be the first to become possessed. If this does not happen within a reasonable length of time, the senior host becomes possessed first and heightens the suggestive effect by dancing in front of the initiate and by making him or her inhale special 'medicines'. It is also worth noting that the host-to-be is seated on a mat right in front of the drummers.[63]

In an article entitled *Percussion and Transition*, Professor Rodney Needham some time ago asked: 'Why is noise that is produced by striking or shaking so widely used in order to communicate with the other world?'.[64] In reply to his question William C. Sturtevant pointed out that:

1. What is required is not percussion per se, but the sort of sound produced especially by drums: Steep-fronted and containing a wide range of audible acoustic frequencies with especially high amplitudes in the lower frequencies.
2. These sounds must be produced with a rhythm near the basic alpha brain wave frequency, i.e. above 7–9 cycles per second.
3. The psychological effects of this sort of auditory stimulation are strengthened by the addition of other stimuli . . . and general stress (e.g. hyperventilation from exertion or fatigue).[65]

Much earlier than this, in an enthusiastic article on the *Ngwayi* drumming practised in Zambia and among the Manyika of Shonaland, A. M. Jones pointed to the 'intoxicating effect' which the crossing of three simultaneously played powerful rhythms has on the listener:

> He can either listen to the main beat of the first and third drums, in which case the beating of the second drum crashes right across it; or he can fix his attention on the main beat of the second drum, in which case the main beat of the first and third drums form a strong cross rhythm which clamours for recognition and is very stimulating.[66]

Among other things, the drums or other suitable sources of rhythm may thus be said to dramatically underscore the situation in which the initiate, and indeed every host to a certain type of stranger-spirit, is conceived to find himself: his spirit has invaded him and 'clamours for recognition'.

Thus the main contribution to advancing the argument is contained in Sturtevant's second and third points. His reference to the importance the alpha brain waves play in effecting dissociation in man, has been borne out by researches among the Tsonga of the Northern Transvaal.[67] Due to their particular geographical situation, the Tsonga have been invaded by undesirable spirits of both Shona and Zulu provenance. These are 'revealed' and exorcised during rituals in which the language, melody, and drum rhythm is indicative of the spirit's supposed origin. Johnston found that the songs which the Tsonga use to effect the exorcism of alleged Shona *chidzimba* spirits call for a drum accompaniment in fast threes, producing a rhythm between 8 and 8.88 cycles per second. The proximity of this rhythm to the basic alpha brain wave strongly suggests that they consciously aim at the psychological manipulation of the subject.[68]

As a result of these researches, one can make an interesting concluding remark. As has been seen, there are, on the one hand, a few 'more serious' stranger-spirits who are believed to endow their hosts with special gifts. Of these, however, it is said that they refuse to 'come out' during a seance which is held for 'dancing spirits'.[69] The majority of stranger-spirits, on the other hand, are said to bestow on their hosts no such special gifts and to possess them only in order to make them dance.[70]

It is quite obvious why this should be so. 'Owners' of the first type of stranger-spirits derive a prestige from the help given to them by their spirits which they would stand to lose if they were to participate in a seance for 'dancing spirits', whose esteem among the people is rather low.[71] Hosts to the second type of stranger-spirits however, derive their gratification mainly from the *ad hoc* pleasures which the dances afford them: the temporary escape from every-day reality; the acting out of stylized roles which nevertheless leave much scope for individuality; the applause from the audience; the abreaction[72] of cooped-up resentments; and the enjoyment of strenuous bodily movement to the accompaniment of powerful rhythms. In this connection, various writers have remarked on the importance of the entertainment aspect in spirit possession dances and on the affinity of these performances to drama and theatre.[73] As will be seen in the next chapter, it is part of the diviner-healer's art to help select the right stranger-spirit for the right patient.

NOTES

1. Singular: *shavi*. Various spellings of this word are to be encountered in the relevant literature. Mine is that given in the *Standard Shona Dictionary*. Hannan (1974), p. 602. Gelfand (1959, 1962) calls *mashavi* 'alien spirits'. 'Stranger spirits' is the term used by Bourdillon (1976), p. 282ff. To avoid confusion, where I quote other sources I will always substitute the spelling as given in the *Standard Shona Dictionary* for whatever other spellings may have been used by the authors.
2. (1969).
3. Bullock (1950), p. 162.
4. Larson (1971), p. 55. Concerning the Shona, Mushoma (1934), p. 31 writes that the *shavi* is a *hombarume* (hunter's spirit) creating the desire to hunt.
5. Gelfand (1962), p. 91-8.
6. Mushoma (1934), p. 31. Douglas (1957), p. 48f says of the Lele that for them 'baboons are unlike other animals in that they will stand up to a man, they experience barrenness, they wash, and they undergo one of the ordeals of initiation'.
7. Beidelman (1973), p. 143 with regard to the Kaguru. Cf. the Shona proverb (Hamutyinei and Plangger [1974], p. 231): 'A baboon is a human being; it knows the art of clandestine stealing'. Similarly, of the Swazi, Kuper (1973), p. 620.
8. Bourdillon (1976), p. 286.
9. von Sicard (1952), pp. 149f. 153, 155, 164. Cf. Hannan (1974), p. 602.
10. Mushoma (1934), p. 31.

11. Cf. Hugo (1935); Gelfand (1959), p. 125; (1962), p. 87.
12. Cf., e.g., Southall (1969), p. 233; Buxton (1973), p. 66; Goody (1962), p. 116f.
13. Bourdillon (1976), p. 287. Cf. Beattie (1969), p. 161.
14. Lienhardt (1961), pp. 163f, 169. With regard to the Mandari, neighbours of the Dinka, cf. Buxton (1973), pp. 64, 66, 105.
15. Colson (1969), p. 78.
16. Note the etymological closeness between *sabe* and *shavi. Masabe* is the plural of *sabe*.
17. Colson (1969), pp. 80; 83-6.
18. Ranger (1975), p. 7. Cf. also Horton (1971), p. 102.
19. Schoffeleers (1975), p. 14.
20. Southall (1969), p. 270.
21. Bullock (1950), p. 159.
22. Stayt (1931), p. 302.
23. Lee (1969), p. 130ff; Sibisi (1975), *passim*. Into this time fell the beginning of the large-scale labour migration to the centres of mining and industry in Southern Africa: cf. Vambe (1972), pp. 223-9; Ranger (1968), p. 161f; Houghton (1974), pp. 402-8. More about the bridge-function of the Tsonga between the Shona and the Zulu/Swazi at the end of this chapter. Lee refers to the Tsonga, using Junod's spelling. Tsonga is the spelling recommended by Van Warmelo (1974), p. 69.
24. Cf. Lee (1969), pp. 130ff; Stayt (1931), p. 303; Southall (1969), p. 263; Walker (1972), p. 95; Colson (1976), *passim*.
25. Aquina (1967), p. 204 says that some diviner-healers in Karangaland specialize in exorcising *shavi*, but they do not form a distinct class.
26. Cf. Lewis (1971), p. 32.
27. Gelfand (1959), pp. 123f, 140f, 149.
28. Gelfand (1962), p. 88f.
29. Gelfand (1959), pp. 126-8.
30. Cf. Bourdillon (1976), p. 285f; (1972a), p. 290. However, speaking of *shavi* cults among the Korekore, the last source says that no male hosts of high social standing participate in them: cf. *loc. cit.*
31. Garbett (1963), p. 106f.
32. Cf. Lewis (1971) and Beattie & Middleton (1969), *passim*.
33. Cf. Bourdillon (1972a), p. 303f.
34. Cf. Bourdillon (1976), p. 287; Gelfand (1962), p. 103-105.
35. Bourdillon (1972), p. 296; (1976), p. 287; Gelfand (1962), p. 102.
36. Cf. Bourdillon (1972a), pp. 292f, 294; Lewis (1971), p. 32 and *passim*.
37. Gelfand (1959), p. 123.
38. For a more detailed description, cf. Bourdillon (1972a), p. 292.
39. Colson (1969), p. 80-2.
40. Bhebe (1973), p. 51. For a similar status-enhancing guild elsewhere, cf. Shorter (1970), p. 117. Also among the Kalanga, and similar to the *shumba* cult, there is the *zenge* possession cult described by Werbner (1964).
41. Bourdillon (1976), p. 283.
42. Bourdillon (1976), p. 284. Concerning the Venda, Stayt (1931), p. 307f. wrote that among them a host to one of the *zwilombo* spirits of Karanga origin

is a valuable asset to a kraal, as she attracts visitors and presents. Powerful chiefs, therefore, like to have a *molombo* dancer among their wives.

43. Gomm (1975), p. 538. At the same time, one ought to qualify Gomm's claim that peripheral spirit possession 'belongs to a class of social phenomena in which someone can be presented as bargaining for something . . . on the basis of what he lacks, is about to lose, suffers from, or is deficient in': *ibid.*, p. 536.
44. Bourdillon (1972a), p. 306.
45. *Ibid.*, p. 307.
46. Bourdillon (1976), p. 285, 287.
47. Bourdillon (1976), p. 283.
48. In some other African societies the metaphor employed is that the host is the horse and the 'power' is the rider: cf., e.g. Beattie and Middleton (1969), pp. 20, 51, 226, 244, 249.
49. Cf. Beattie and Middleton (1969), *passim.*
50. Bourdillon (1972a), p. 295. For the Shona in general, cf. *ibid.* (1976), p. 280, and Gelfand (1962), p. 169. Cf. also Beattie (1969), p. 165; Middleton (1969) p 225 f.
51. Cf. Bourdillon (1976), pp. 274-88; Gelfand (1962), pp. 84-98, 100-5.
52. I did not mention this possibility before, as it adds nothing to my argument concerning the *midzimu.* For detailed descriptions, cf. Bourdillon (1976), pp. 263, 279-82; Gelfand (1959), pp. 75, 92-4, 96-8.
53. Cf. the photographs facing pp. 40, 57, 72f, 136, in Gelfand (1959), and pp. 86 102, 118, in Gelfand (1962). Bullock (1927), p. 162 complains about the 'regrettable tendency shown by the younger generation of the Mashona' to imitate the *shavi* cult by talking in strange tongues and by decking themselves in the special calico demanded by the *zwipuna* spirits. Similar observations have been made by Southall (1969), pp. 240, 243 and Tanner (1969), p. 279. Among the Xhosa in South Africa, I have often witnessed the relish with which children mimic various categories of 'possessed' people during their school 'concerts'. Cf. also the telling photographs of boys imitating a trance in L. Marshall (1962), Plate 14.
54. First indications that a person is to become host to a spirit often occur in early childhood. A person is said to become a medium (normally) while in his late teens or early twenties: Bourdillon (1976), pp. 277, 283. Gelfand (1964), p. 82 writes that he has seen girls of twelve and fourteen entering into a trance during a *shavi* ceremony.
55. Cf. Chapter three, subsection *Authority and ownership.*
56. Cf. Bourdillon (1976), p. 277. Gelfand (1959), p. 152.
57. Cf., e.g., Buxton (1973), pp. 104-13. With a view to my later discussion of Shona Independent Churches it is interesting to note that this is particularly true of those areas in which Islam has gained a considerable foothold among the population: cf., e.g. Shack (1971), Gray (1969), Lewis (1969).
58. Cf., e.g. Horton (1969), p. 28ff; Colson (1969, p. 78ff; Lee (1969), pp. 130-3; Southall (1969), p. 258-71; Lewis (1971), pp. 66-99. In Bunyoro both accomodation and exorcism are possible, though only few *mbandwa* spirits can be destroyed and exorcism is regarded as a highly dangerous operation: Beattie

(1969), p. 164. Shona diviner-healers may also attempt to exorcise spirits, but the chances of success are as limited as in Bunyoro, as will be seen in chapter five.

59. Cf., e.g. Bourdillon (1976), p. 278; Gelfand (1962), p. 85.
60. Beattie and Middleton (1969), pp. 36, 51, 88, 334-8.
61. Bourdillon (1976), p. 275.
62. Bourdillon, pp. 279, 285. Similarly Colson (1969), p. 79 of the Tonga. The effect of the exciting rhythmic music on 'heating the spirit' is also emphasized in Beattie and Middleton (1969): cf., e.g., pp. 7, 23, 59, 79ff, 87, 244.
63. Cf. Bourdillon (1972a), pp. 291, 295; Gelfand (1962), p. 101.
64. Needham (1967), p. 607.
65. Sturtevant (1968), p. 133f.
66. Jones (1943), p. 17f. Gelfand (1959), pp. 139, 148 and, (1962), pp. 94 and plate 7, foll. p. 86; 90, 92, 98, mentions the use of *three* drums in connection with several *shavi* cults.
67. Cf. Johnston (1975).
68. *Ibid.*, pp. 784, 787 (his spelling is *xidzimba*). The auditory driving is further enhanced through the use of drugs. For a short overview of research into the correlation between alpha brain waves and dissociated states of mind, cf. Watson (1974), p. 222ff. Verger (1969), p. 59 says of the Nago-Yoruba and Fon of Dahomey, that among them each *orisha* or *vodun* has his own rhythm to which he is sensitized in the course of initiation. In Dutch Guiana and Dahomey, cult members 'fear hearing their rhythms because they are afraid that they will begin to dance and never be able to stop': Walker (1972), p. 76. Bourdillon (1976), p. 285 and Gelfand (1959). pp. 125, 132, 136, 138, 142, 147, 149, 151, (1962), pp. 85, 89, 90, 91f, 101f, 105, (1964), p. 40, also state that *shavi* cults have characteristic rhythms, though these are not always produced by drums. Gelfand (1964), p. 81 remarks on the connection between the tempo of the music and the host's entrance into the trance state.
69. Bourdillon (1976), p. 287.
70. *Ibid.*, p. 282f.
71. *Ibid.*, pp. 285-7.
72. *Pace* Lewis (1971), p. 195.
73. Bourdillon (1972a), p. 303f; (1976), pp. 275, 285, 288; Lewis (1971), p. 195; Walker (1972), pp. 65, 86. Ranger (1975a) describes an East African dance organization which did not originate from a spirit possession cult. However, its function was akin to the latter and its influence stretched well into Rhodesia. Cf. also Beattie (1977).

CHAPTER 5

WIELDERS OF POWER TO HARM AND TO PROTECT: THE WITCH AND THE DIVINER-HEALER

For the Shona the important factor in witchcraft, as well as in divining and healing, is the special power with which those engaging in these crafts are believed to be endowed. Just as in the preceding chapter the Shona have been seen to explain the extraordinary skills a person may develop as a hunter or drummer, by saying that these skills are given to him by his stranger-spirit, (*shavi*), the most powerful – and therefore most feared witches are also thought to be the hosts of specific witchcraft *shavi* (*shavi ro uroyi*) who give them their ultra-human powers to do evil. Witches (*varoyi*) say that their witchcraft *shavi* urge them to kill and to eat people, very often their own children, or to cause grievous harm to both kinsfolk and unrelated neighbours, either in order to increase the power of their charms or for the sheer pleasure of doing evil.[1]

Witchcraft *shavi* are generally believed to be inherited, and women in particular, are associated with inherited witchcraft. There is, therefore, some superficial similarity between the Shona and those African societies in which there are 'two kinds of witchcraft: psychic power used by women, and external magic used by men in furtherance of acceptable political objectives', to which Mary Douglas refers in her introduction to *Witchcraft Confessions & Accusations*.[2] Bourdillon, for example, says with regard to the Korekore chiefdoms of Diwa and Chiruja, where diviners, herbalists, and spirit mediums abound 'men have the authority and women are feared as the most powerful witches'.[3] A female witch is believed to inherit the witchcraft *shavi* from her mother or from some other maternal ancestor, and upon her death she in turn is thought to pass her *shavi* on to a daughter or grand-daughter. Therefore, the children of reputed witches are always suspected of being witches themselves.[4]

Of the Karanga, it is said that in their belief a witchcraft *shavi* is never passed on in the agnatic line, so that the taint of witchcraft cannot adhere permanently to any particular patrilineage, 'but touches many patrilineages on their periphery', because 'witchcraft is passed on from one married woman to another, all of whom have lived as stranger-wives away from their kinsfolk'.[5] The solidarity of the lineage is thus maintained at the expense of non-lineage members. The same principle of solidarity, however, assures the accused female witch of sympathy in her own lineage. If she is divorced by her husband, her own kinsmen will accept her, because they do not believe the charges levelled against her by what, to them, are the 'outsiders'. Since a witchcraft *shavi* tends to become permanently associated with a particular line of mothers and daughters, some writers have even spoken of 'the *mudzimu* [ancestral spirit] which possesses a witch [and] is thought to be the driving force behind her'.[6]

Affinity with spirit possession

Professor I. M. Lewis makes the important observation that 'it has long been recognized that unsolicited possession by malevolent, capricious spirits may serve as an explanation of illness and affliction in much the same fashion as witchcraft and sorcery'.[7] Although he does not include the Shona among the few scattered African examples he uses in his essay, his observation also applies to the Shona. According to Lewis, spirit possession and witchcraft represent different strategies of attack, the first being of a milder and less disruptive type than the second. It is restricted to the domestic domain, where female dependants can use it against their superiors. By contrast, witchcraft accusations tend to operate in a wider sphere of interaction and can be utilized as a strategy of attack in general contexts of hostility, particularly between equals. A further interesting point made by Lewis is that persons who succumb to the attack of a peripheral spirit may graduate to become spirit-possessed diviners, once they have learned, through their membership of a spirit possession cult, how to control the spirit. Having obtained control over their spirit familiar through accommodating themselves to it, they may eventually make use of its power for launching witchcraft attacks against other people. At any rate, the fact that, as diviners, they assume a more active role, can readily lead to witchcraft accusations being levelled against them.[8] As will be seen, this is precisely what is happening to many Shona diviners, particularly those to whom Gelfand refers as 'black *nganga*', i.e. diviner-healers who supply their clients with 'medicines' intended for harmful purposes.[9]

There are a number of further points which emphasize the affinity between spirit possession and witchcraft in Shona theory and practice. Firstly, Bourdillon states that a person to whom it has been revealed by divination that he is wanted as a host to a witchcraft *shavi* may try to get rid of the attacking spirit. This may be done either by attempting to have it transferred into an animal by a diviner-healer, the animal then being driven into the wild; or by approaching a prophet of one of the Independent Churches to have it exorcised. However,

> since a spirit can rarely be thwarted from achieving its purpose, these attempts may fail and *the person concerned may in the end accept the spirit and become a witch*, this being in her view a lesser evil than the trouble caused by the spirit.
> Sometimes a witch is persuaded to practice by other witches and . . . normally it is assumed that the witches who teach the new witch her craft also provide her with a helping spirit.[10]

The words in this text which I have italicized for emphasis will immediately call to mind what has been said in chapter four with regard to the host of a stranger-spirit 'coming to terms' with this spirit.[11] Another similarity between witchcraft and spirit possession is apparent from the second part of this quotation. Witches, particularly if they are female, tend to form covens, although the number of their members would seem never to be as large as that of an ordinary *shavi* possession cult. This is not surprising, in view of the weirdness of their real or imputed acts and the abhorrence most people have of the same. However, there exists a small body of confessions which have been made by alleged Shona witches in the course of criminal investigations against them, in each of which two or more women claim to have committed witchcraft together.[12] Crawford points out that such voluntary confessions are attributable most particularly to two categories of women, who intend to enhance their status in their community by surrounding themselves with an aura of fear and respect. They are either young women who are barren or who keep losing their children during pregnancy or soon after birth, and who are therefore looked down upon by their husbands and co-wives; or they are old women past the child-bearing age who, in contrast to their young counterparts, are relatively secure and who may use the confessions to establish themselves in the eyes of the public as particularly powerful ('black') diviner-healers.[13]

Thirdly, characteristic traits of spirit possession may be advantageously combined with accusations of witchcraft. This is illustrated by a case in which the accused man was a Korekore school teacher named Nyakadzino. He had intended to marry the daughter of Mangava, a Salvation Army Captain, but was rejected. Thereupon Nyakadzino, two other teachers, and twelve children became 'mentally deranged'. The afflicted clung together and said that if they were to be separated they would die:

> They spoke as one . . . Accused . . . said his arm was stiff and only after an attempt by the children could he bend it . . . Accused told me he would be unable to remember what had happened during his period of mental sickness when he recovered. Whilst I was there an Apostolic prophet arrived and there followed . . . a witch hunt, with prayers and singing. As a result, certain articles were found in and around the school area: . . . a pin covered with black putty bound round with red wool . . . was found buried in the earth beneath the seat occupied by Mangava at school.[14]

In the end Nyakadzino, while in his 'deranged state', attacked Mangava's property and from this action arose the court case against him.

Strategy of attack

Bourdillon writes that accusations of witchcraft are almost always preceded by tension and conflict within the community.[15] One would expect this to be the case if indeed witchcraft allegations are, as Lewis has claimed, a strategy of direct attack which people employ against each other in the context of general hostility. He goes on to say that the accusations can be the result of conflict over succession and leadership, or they may also arise from disagreements over the distribution of family wealth, or any other dispute within the family group. Since conflicts of this nature should not be contested in a public court, but rather solved at the family level, simmering tensions readily erupt in accusations of witchcraft, which provide an excuse for breaking up the group. According to Mary Douglas' scheme, which she presents in her introduction to *Witchcraft Confessions and Accusations*, Shona society should be classified along with others which have a 'witchcraft-dominated cosmology'.[16] Although belief in witchcraft is almost universal among the Shona,[17] both Crawford and Bourdillon would seem unwilling to call Shona cosmology 'witchcraft-dominated'.[18] However, Douglas's assertion that witchcraft allegations in a society which has a 'witchcraft-dominated cosmology' can be traced to a set of definable social niches[19] is supported by the Shona material, as Bourdillon's reference to endemic areas of conflict indicates.[20]

Although the Shona view of the world can hardly be said to be 'witchcraft-dominated', it is nevertheless, in the words of Crawford, 'not easy to think . . . of any other type of accusation which can so readily embitter social relations dna increase social tension.'[21] Such allegations frequently lead to assault, resulting in serious bodily harm or death; sometimes they have caused the accused to commit suicide; often they result from marriage difficulties and provide the alibi for a desired divorce; at the very least, they always lead to the witch's being avoided, while more often the accused is forced to leave the community for good. The severity of the negative sanctions traditionally imposed on persons accused and convicted of witchcraft necessarily leads one to ask what it is precisely that Shona people conceptualize under the guise of the witchcraft paradigm. I made one important observation, when I noted that the most powerful witches are thought to be possessed by specific witchcraft *shavi*. However, in the same way in which *shavi* function, among other things, as conceptualizations of what would otherwise remain for the Shona inexplicable powers experienced in their daily lives, Bourdillon writes that their 'witchcraft beliefs provide a rationalization of suspicion and tension'. He goes on to say:

> Although to the modern western mind these beliefs may appear as irrational superstitions which can divide a community or bring unjust opprobrium on persons who have fallen victim to vague suspicion, to the Shona mind the beliefs are based on a real experience of the world about them. Evil, troubles, sickness are undeniable and demand explanation; social tension and conflict frequently arise, a clear sign that some people are careless of the good of their neighbours; bad people are to be found and appear at times to have their way.[22]

Evidence from witchcraft trials shows that there are in fact people who are prepared to commit serious crimes, including murder, in order to procure the ingredients for powerful 'medicines' which they use for such disparate aims as securing a lover's affection and procuring success in business or gambling. Sometimes it is found that 'medicines' used for 'bewitching' a person do in fact contain physical poisons, such as arsenic from cattle-dip.[23] Even committing incest for the sake of improving the harvest has been the cause of trials for alleged witchcraft.[24] I have noted already that persons will occasionally also make unsolicited confessions of witchcraft. All this massive evidence has come to light in spite of the fact that the Witchcraft Suppression Act threatens with heavy sentences anyone who is convicted of imputing witchcraft activity to others, and that for this reason relatively few witchcraft cases are tried in court. One may assume, therefore, that the ordinary Shona has more than ample evidence on which to base his conviction that there are persons who delight in horrific deeds and who have extraordinary powers.

Paradigm of anti-social behaviour

'A witch is a chief,' says a Shona proverb,[25] which is commonly quoted by mourners who suspect that witchcraft has caused their bereavement. It clearly implies the commonsense explanation that 'a witch is practically above the law . . . and behaves like a despotic chief'.[26] However, it would seem also to echo many of the stories known to the Shona which specifically associate chiefs, as 'owners of the land', with the possession of powerful protective charms; and contenders for the chiefly office with the use of witchcraft against their potential rivals.[27] In this connection I mentioned in chapter two that the mythological founder-ancestors of Shona chiefdoms often did 'what others were afraid to do', such as, for example, committing incest with a sister, and openly disregarding the social order, thereby acquiring unusual power over the physical order. The same rationale stands behind the way in which the present-day Shona conceptualize their witches: their conduct is an inversion of that which is approved of by society, and this disregard for generally accepted norms gives them their extraordinary power to harm their fellow men. Thus, for example, they go about their nefarious deeds during the night, when normal people are asleep; they do so naked, and often use objectionable animals such as hyenas as their mounts. In addition they have a perverted predilection for human flesh.

The more violently a witch profanes and rejects the values of society, the greater becomes his or her power over physical nature.[28] Once this principle has been accepted, imagination is free to add colour to the picture:[29] witches can leave their body asleep in their dwellings while they fly around on their nocturnal activities, covering great distances; they exhume recently buried bodies and deprive them of those parts which they need as ingredients for their 'medicines', without leaving any trace of injury; they are supposed to be able to arouse an avenging spirit to help them achieve their ends, and they can send down lightning upon any of their victims. To the Shona, therefore, 'witchcraft is the paradigm of all evil and anti-social behaviour',[30] and they readily suspect it to be at work wherever something unpleasant or very much out of the ordinary happens, especially if it is accompanied by discord in the community.[31]

When somebody dies, particularly if it is a young person, witchcraft is nearly always presumed to be the cause. When a woman suffers a number of stillbirths, or when several of her children die at birth or soon afterwards, she is presumed to be a witch. Serious and persistent illness, as well as the sudden and sharp pains of rheumatic disorders, are associated with witchcraft, especially if there are quarrels within the family. If snakes or owls and other animals associated with witchcraft visit a homestead, they are presumed to be a witch's familiars (*zwidoma*) whom she has sent there to hide her dangerous 'medicine' in and around the houses where it will cause harm to both man and beast. A person's great and apparently inexplicable success in comparison with those around him is liable to be attributed to his use of witchcraft. Likewise, a very arrogant person who appears to fear nobody and who readily speaks his (or, more commonly, her) mind is assumed to be confident because of his mastery over the powers of witchcraft. Indeed, 'any kind of power may be associated with witchcraft, since power includes power to harm'.[32]

Crawford has taken to task those social anthropologists who have put 'too much emphasis on the allegation of wizardry as being the reflection of social tension', because in his view this 'leads one into the error of supposing that an allegation is a mere symptom of social malaise and is not, in itself, a dynamic force in the community'.[33] He points out that an allegation of witchcraft may originally reflect a state of emnity between two persons. However, once the accusation has been made, 'a chain of events may be set in motion which can lead rapidly to the most unforeseen and often tragic consequences'. For him, therefore, the allegation of witchcraft is primarily a device used to involve the whole community emotionally in what would otherwise be merely a private quarrel. The device is used by one or both of the conflicting parties to marshal public opinion against a particular person. Once that person has been labelled a witch, nobody is supposed to be able to remain neutral towards the conflict since witchcraft is the worst thinkable anti-social evil. It will be noticed that Crawford's argument essentially anticipates what Lewis has said about witchcraft as being a 'strategy of attack'.

Although the witch's use of his or her power to influence people's fortune in an adverse manner bears a certain similarity to the action of the territorial and ancestral spirits, who are also thought to bring bad luck upon those under their sway, the manner in which they do so and the reason for which they do so differ radically from those ascribed to the witch. Apart from the witch's typical link with what is abnormal and anti-social, what makes witchcraft attacks so frightening is that they are utterly evil, because they are directed against innocent people and spring from the basest human motives, such as envy, jealousy, greed, hatred, and the craving for power. A person who has been attacked by a witch is believed to be almost completely at the mercy of his attacker. Only sometimes may a diviner-healer be able to neutralize the effect of the witch's 'medicines'. Often it is thought that the only way for the attacked victim to get a reprieve is by asking the witch to withdraw his or her evil influence.[34] In contrast, although the Shona feel that the territorial and ancestral spirits at times afflict their dependants too severely, or for no apparent reason, they are believed to punish out of concern for the person's well-being, and there are recognized ways in which they can be appeased.

In chapter two it was seen that the White settlers in Rhodesia are felt to have 'spoilt' the land. They have broken the power of the chiefs, and destroyed the relationship which had traditionally existed between the land and the chief. An early indictment of this kind mentions a second factor which is held responsible for the 'spoiling' of the land, namely the unchecked activity of the witch:

> The English have taken the country, we don't get sufficient food, the English won't allow us to drive out witches; before the English came we were allowed to drive out all evil-doing persons and consequently did not die at the high rate that we do now.[35]

The same complaint is still prevalent among contemporary Shona. It gains support from their experience that some people do in fact make use of the freedom now given to witches by acquiring a reputation for witchcraft and then holding considerable sway over the community in which they reside through the judicious use of thinly veiled threats.[36]

I have already mentioned that women in particular may enhance their status in the community by acquiring the aura of fear and respect which surrounds a known witch.[37] Therefore the belief in witchcraft remains 'very real and very strong; many hamlets in the rural areas and numerous urban homes are protected against witchcraft by ritual and charms, and persons often wear charms and take medicines to protect themselves from witchcraft'.[38]

For this reason, traditional attempts to eradicate witchcraft by curing the witches of their evil ways and by destroying their 'medicines' have retained their attraction among the people to this day, especially in the rural areas. Periodically, movements for the eradication of witchcraft have made their appearance, although they all seem to have originated outside of the country.[39] It seems, however, that nowadays the function of these traditional movements is increasingly taken over by the Independent Churches, as will be seen in chapters six, seven and eight.

The diviner-healer

A corollary of the widespread belief in witchcraft is the continued presence of a large number of diviner-healers[40] throughout Shonaland. In the towns and cities their number is even said to be on the increase, because 'the tensions and uncertainties of modern life readily provoke suspicions of invisible forces, especially of witchcraft, and few Shona can resist going to a diviner when things are going badly'.[41] The remainder of this chapter will take a closer look at the *nganga*. It will deal particularly with the manner of his calling, and with those aspects of his divining and healing, and other practices, which are relevant to the theme expressed in the title of the present chapter. There will also be occasion to note the obvious parallels existing between the witch and the *nganga*, which account for the ambiguous status he has in the eyes of the people.

There are many Shona who have an elementary knowledge about

'throwing the bones' and who may, therefore, occasionally be asked by neighbours or relatives to divine for them in minor troubles and anxieties. The professional *nganga*, however, needs a special divining or healing spirit as help, just as the 'real' witch is distinguished by her being host to a special witchcraft *shavi*. Most *nganga* use one of the following two divining methods, or, sometimes, both together: they either consult their divining dice (*hakata*),[42] or deal with clients while 'possessed' by their healing spirit. In the second method the aid given the *nganga* by the 'spirit world' is obvious, but his ability to throw the dice and to interpret their fall is no less attributed to direction by his in-dwelling stranger-spirit.

Calling

As is the case with the prospective witch,[43] the future *nganga* usually finds out about his stranger-spirit's wish to enter into him, in his dreams. In the majority of the twenty-six case-histories recorded by Gelfand,[44] the *nganga* had received their calling to the profession through persistent dreams about their endowment with divining and healing skills in the future. Most of these men and women have had relatives who had themselves been *nganga*. The ancestral spirit of their dead relatives would appear to them in their dreams and inform them of his wish to continue his craft through them. As Gelfand puts it, the healing *shavi*, who is by definition a 'stranger-spirit', becomes as it were 'naturalized'.[45] Once it has taken possession of a medium in a particular family, it selects all its subsequent mediums from that family. After a few generations, the 'stranger-spirit' has become part of the family and is regarded almost in the same light as an ancestral spirit.[46] Often the candidates suffer in addition from serious illness and many of them go through periods of mental confusion.[47]

As in all other cases where a person is presumed to be called to become the host of an ultra-human power, the final verdict has to come from a *nganga*. Frequently the family or the afflicted person himself do not trust the verdict of just one diviner, and several others may be consulted before it is finally accepted. A ceremony is then held in honour of the invading stranger-spirit. In the case of women, barrenness or an inability to raise children is often interpreted as a sign that a divining or healing *shavi* desires to enter into them. Accepting such a *shavi* holds out for them the possibility of winning a double prize: frequently they are promised that if they accept the *shavi* they will be able to raise children; however, even if they should remain childless, their newly gained status as *nganga* more than compensates them for the disgrace which the Shona public attaches to a woman's childlessness.

Usually there follows a period during which the designated *nganga* is apprenticed to a renowned practising diviner-healer in order to be taught the craft by him, upon payment of a fee. Others, however, practise without having received such formal instruction. They maintain that they have been taught solely by their spirits. Some of them claim that they are told in their dreams which treatment to use in each individual case, and where to find the necessary remedial plants. There are also stories told of particularly powerful *nganga* who are said to have received their training from the spirits while they were living with them for a few days at the bottom of a pool. It will be noticed how these stories echo the theme of the mythological pool, which has been discussed in chapter three.[48] Most *nganga* begin their practice only after they have gone through a final ceremony, similar to the kind which is held when new members are initiated into *shavi* cults. Its purpose is to allow them to give proof that they are really possessed by a *shavi* and to put the stamp of public approval upon their subsequent work as diviners and healers. Many *nganga* specialize in a particular skill, such as divining the cause of death, 'sucking' illnesses from their clients' bodies, treating infertility and impotence, epilepsy,[49] as well as driving out avenging spirits and doctoring homesteads against lightning and witches.

I have indicated above that the *nganga's* help is sought in a variety of situations. This fact is clearly reflected in the high number of court cases which have been brought against *nganga* under the Witchcraft Suppression Act. Out of the 103 cases recorded by Crawford, 77 involved Shona people as defendants and plaintiffs. In 41 of the Shona cases the accused was a *nganga* or a prophet of one of the Independent Churches under trial for imputing witchcraft to others.[50]

Divination

Nganga who divine without the use of dice enter into a ritual trance at the beginning of each divining session and any pronouncements they make while they are in this state are supposed to come directly from the stranger spirit(s) to whom they are host. Like any other 'possessed' host, a diviner of this type needs an assistant who receives the fee and helps him to enter into the trance state. Throughout the session, the assistant also serves as his interpreter, since the in-dwelling stranger spirits are often made to appear to be speaking in foreign dialects. Not surprisingly, the assistant is usually the *nganga's* spouse or someone related to him.[51]

According to Crawford, the method of divining with the use of various types of dice is the most widespread among the Shona.[52] The process of divination consists of putting questions to the dice before the throw and then reading the answer from the position in which they land on the ground. Depending on the question which has been asked, the answer is either 'fortunate' or 'unfortunate', or 'yes' or 'no'. While the method bears some similarity to the Zande poison oracle,[53] the Shona divining dice are a far more subtle instrument than the latter. The most widespread form consists of sets of four oblong pieces of wood, ivory or bone, each of which has a pattern carved on one side. Thus, depending on which of the dice fall with the carved side uppermost, there are sixteen possible throws, each of which has its proper name and a range of meanings. Many *nganga*, however, throw four or eight sets of dice together at a time, thereby increasing the number of possible combinations and complicating the interpretation.

It may seem that divination by means of the divining dice is basically a purely mechanical process, although it is quite obvious that the diviner, through skilful manipulation of the dice, can influence the outcome of the different throws. I have stressed, however, that in the eyes of the people who are seeking the diviner's help it is his *shavi* who correctly enables him to read the message conveyed by the dice. In fact, the divining dice themselves are said to 'see the truth' and to 'speak' through the manner in which they arrange themselves in the course of each throw. Therefore, a *nganga* cannot use a new set of dice before he has 'medicated' them through an elaborate ritual designed to take away their 'blindness'. Gelfand says that even afterwards many *nganga* treat their dice in a ritual manner each time before a divining session.[54]

When speaking of the possibility that the *nganga* manipulates his divining dice in order to attain a desirable verdict, I do not mean to cast doubt on the sincerity of the majority of the Shona diviners.

Crawford says he has been impressed with the diviners' obvious sincerity and belief in their craft. Yet he also speaks of the necessity for the *nganga* to avoid an unpopular verdict. He believes that 'this is . . . to some extent an unconscious process',[55] and Bourdillon notes that during divination, a diviner repeatedly gathers up and throws his dice as he asks questions of the clients who have come to consult him.

> Even when the clients are able carefully to consider all the throws, it is extremely rare that the combination of the different sets used in a divination provides an unequivocal answer to the questions asked of the dice. The diviner must . . . utilize his prior knowledge of the background to the case in question if he has any, and whatever a skilled diviner may infer from an astute assessment of the reactions of his clients during the course of the divination.[56]

In this connection it is interesting to note that in the beginning of the divining session, the *nganga* usually hands his dice to his client and lets him make the first throw. This is said to have the purpose of allowing him to 'make contact' with the dice. Only afterwards does the diviner begin to throw them himself and to interpret the answers they give to his questions.[57] Particularly where it happens that a party consults a *nganga* in order to find out which of them is responsible for a suspected case of witchcraft, the diviner may cause each of the arrivals to throw the dice and then pronounce his verdict, apparently basing it on the manner in which the dice fell in each individual throw.[58]

As I have mentioned above, although people usually accept the result of a *nganga's* divination of their case, they are free to reject it. They are particularly likely to do so if it is quite contrary to their own expectations. Crawford says that very frequently the diviner's verdict is only accepted after he gets angry with his clients or tells them that if they do not want to believe him they should consult someone else.[59] It is even admitted that professional diviners of considerable repute may err in a particular divination. However, this does not shake the Shona's faith in the divination process itself, as to them its reliability is ultimately guaranteed by the intervention of the *nganga*'s *shavi*. In the case of witchcraft suspicions, the only way in which a person could traditionally prove his innocence beyond any doubt was by means of subjecting himself to various kinds of ordeals. Even today it happens occasionally that women seek an ordeal of their own volition as a means of freeing themselves from an intolerable situation.[60]

From all this it is clear that the *nganga's* diagnosis of a case is far from depending solely on the verdict returned by the fall of his dice. It is even more obvious where he arrives at his verdict without the use of mechanical devices, divining while he is in a ritual state of trance. Bourdillon writes that 'ideally, before making a diagnosis, a diviner shares with his client every detail of the client's private life: his family relations, his fears, his enemies, his past history', and he notes the obvious parallels between this method and modern psychotherapy:

> In this process, any qualms of conscience are likely to be brought into the open and might suggest possible causes of the troubles of the client. Any serious immoral or anti-social act is likely to suggest itself as the cause of misfortune, in which case confession and the consequent release of tension is usually a step towards overcoming the problem.[61]

It is customary, first for the *nganga* to 'go round the client's spirits'; i.e. he goes with the client through the names of all the ancestral spirits who might have caused the misfortune against which he seeks the diviner's help.[62] His questions aim at finding out whether the client feels uneasy about having neglected some ritual obligation towards his ancestral spirits or some other spirit, such as a stranger-spirit which may be associated with the family. If the client's relationship with the spirits who are thought to control his world is found to be in order, the *nganga* next turns his attention to the possible existence of tensions between the inquirer and other people. Confession is particularly important in cases where the rights or the property of others have been infringed by the inquirer or any other member of his family. One would expect this to be the case in view of the importance which authority and property have in Shona cosmology, as has been pointed out in chapter three. From the first part of this present chapter it is also clear that the discovery of any tensions in the client's social relationships permits the *nganga* to suspect that witchcraft has been used against him by those who bear him a grudge. A consultation may, therefore, go on for as long as three or four hours, because the family is naturally anxious to be quite sure that the nganga's diagnosis is correct.[63]

I have also mentioned that, apart from his astute assessment of the reactions of his clients during the course of the divination, the diviner utilizes whatever prior knowledge he may be able to gain about the background to the case in question. This is no problem if he resides in or near the community from which the consulting party comes. Bourdillon notes that in cases where clients are from far away and therefore unknown to the diviner, the latter would usually provide hospitality to them before making his divination. This would give him and his assistants the opportunity for a certain amount of eaves-dropping.[64] The most sought-after *nganga* are often said to surprise parties coming to them from very distant areas by the exact fore-knowledge they have of the purpose of their visit. Good diviners are even supposed to know the amount of money the party have brought to pay them for their services.[65] Gelfand tells of an instance when he was tremendously impressed by the detailed knowledge which a male *nganga*, whom he had never met before, had about his family, while possessed by a female healing *shavi*. Later, however, he found out that as soon as he had arrived in the diviner's village, the latter had elicited all of this esoteric information from the nursing orderly who had accompanied Gelfand on the trip.[66]

There is good reason for my having treated so extensively the various aspects of the divining methods used by the *nganga*. As Bourdillon points out, 'there are two distinct aspects to the Shona reaction to personal misfortune'.[67] Of these, the first, concerned with determining its 'ultimate spiritual cause' and with appeasing it, today remains the most conspicuous in the *nganga's* daily practice. In western eyes this has earned him the ill-chosen epithet 'witchdoctor'. The second aspect concerns the treatment of disorders which have been caused by the misfortune. Where the disorder is a case of sickness, it is today increasingly handled by western medicine and thus tends to be taken away from the *nganga's* immediate sphere of responsibility. However, for most of the Shona only the *nganga* can give the answer to questions such as these: 'What has caused a particular person's misfortune; why should it have picked him and not somebody else; why did it strike him at this particular time';[68] and (particularly where the harm done does not consist of illness, or where the illness is of a type which, it is thought, cannot be cured by the Whites)[69] 'what can be done to destroy or neutralize its evil influence?'.

The traditional Shona approach to misfortune is very well reflected in two of their proverbs. The first of these expresses their firm belief in the rationality of their quest for the cause of evil events: 'Something has killed the owl; it cannot be the wind'. The second indicates where a man is to seek help after he or his family has been struck by misfortune: 'Bad luck befalls a dog; a person's is brought to a diviner'.[70] In a cosmology which, I have argued, conceptualizes ancestral spirits, stranger-spirits, and witches as those agents ultimately responsible for any death, illness, or personal misfortune, the *nganga* obviously holds a key position and his divining powers would be even more important to the people than his ability to remedy the effects of attacks sustained from these invisible forces.

'Medicines'

The second aspect of Shona reaction to personal misfortune, namely the treatment of disorders which have allegedly been caused by it, leads to the discussion of the nature of the 'medicines' used by the *nganga*. I have already mentioned that where modern, western-style medical facilities are available to the Shona, they tend to make increasing use of them, in the hope of finding an alleviation of their physical ills. Indeed, *nganga* are said to be happy to send difficult cases to a 'White' hospital. However, while Shona people readily acknowledge that western medicinal treatment can often cure their ills far better and more quickly than treatment administered by their own *nganga*, they also believe that western diagnosis is quite unable to discover the ultimate cause of their illnesses. For this reason, Shona patients treated by a western-trained physician are likely to consult a *nganga* also. Gelfand notes that today there is a 'constant flow of African patients between the doctors of both civilizations'.[71] In accordance with the traditional Black doctor's wider definition of what constitutes a disorder amenable to medicinal treatment, he supplies his patients with 'medicines' and treatment for situations which his western counterpart has to admit lie outside his professional competence. The *nganga's* divination seeks to reveal the powers responsible for any of his clients' ills and misfortunes. Consequently, his treatment is either aimed at appeasing the powers which are held responsible for the disorder (this would be especially true where ancestral and stranger-spirits have been divined as causes of the misfortune) or his treatment consists in launching a counter-attack

against the powers by supplying his clients with 'medicines' designed to break the influence which the powers have gained over them. Thus, as Bourdillon points out in an example, an unsatisfactory wife may blame her laziness on bewitchment, and her husband might feel that his 'bad luck' with her can only be overcome by acquiring from the *nganga* new and more powerful 'medicines'. Or otherwise, bad luck in marriage may be attributed to a stranger-spirit which must be appeased or exorcised[72] with the help of a *nganga*.
It is important to realize that 'medicines' are to be seen as belonging to the same category as those powers which the Shona visualize as spirits. As I have mentioned in chapter two, the reason why the first missionaries were often eagerly welcomed in Shonaland was partly that local chiefs were anxious to gain access to the 'medicines' which they thought were responsible for the foreigners' admired marksmanship. Gelfand notes that another avenue along which the missionaries were able to make friendly contact with the Blacks was their knowledge of medical treatment:

> Knowing that the missionaries were ministers of religion, the Africans believed that they were endowed with healing properties in the same way that they believed their own *nganga* were. The sick were very anxious to be cured by the missionary, who might have better remedies than their own doctors.[73]

Ambiguity: Power to protect and power to harm

In the same way in which a stranger-spirit is thought by the Shona to be responsible for a person's particular skills, 'medicines' which have been prepared by the *nganga* under the direction of his healing spirit are believed to give to their users the power to free themselves from whatever inconviences they may suffer. Thus *nganga* provide 'medicines' designed to ensure their user a good harvest (*divisi*); or success in school examinations, in finding and holding a good job, in gambling, in winning a person's affections, etc.[74] However, the fact that both 'good' and 'bad' *divisi* are available for money from *nganga*[75] proves the correctness of placing the diviner-healer, as wielder of power, on one and the same continuum as the witch. Gelfand's reference to the Shona belief in the existence of 'black *nganga*' shows that the people are quite aware of the fact that some of their *nganga* may actually be sorcerers,[76] just as Bourdillon has encountered 'fringe cases' among Korekore mediums of territorial spirits who were more like a *nganga*, because they earned a living by divining and healing, and their influence among the people was mainly based on their practice of these crafts.[77]

However, although there are unscrupulous *nganga*, particularly in towns, where some of them may become quite rich, most diviner-healers are said to genuinely desire to assist the people in their troubles and to use their powers for their clients' protection against the powers of evil.

Witchcraft is often spoken of by the Shona as a form of 'poisoning'.[78] Gelfand says that the witch sends her bewitching 'medicines' like an 'invisible missile' against her victims.[79] She may also administer them in person during her nocturnal travels. She then either forces the victim to drink the 'poison' while he is in his sleep,[80] or more commonly, makes tiny incisions on the limbs she wishes to affect, into which she rubs *chitsinga* or *chipotswa*. These 'medicines' cause severe pain to the victim and are often accompanied by a swelling of the limb.[81] Gelfand says further that since the conditions believed to be caused by a witch's *chitsinga* or *chipotswa* are so common and so refractory to treatment, no *nganga* can afford not to have treatment to remove these 'medicines' from a client's body. Usually this consists of his making incisions in the painful area and sucking the evil matter out of these cuts, either with his mouth or by using a cupping horn. The extracted 'medicine' is then shown to the client – usually it is a small worm – and disposed of in a ritual manner.[82]

As briefly mentioned above, *nganga* may be called upon to attempt the cure of a person believed to be possessed by a witchcraft *shavi*.[83] One method in which this may be done, and which has been described by Gelfand,[84] is interesting, because it will be found in chapter eight that it plays an important role in the exorcisms carried out by some of the Independent Churches. During this treatment, the witch has to drink a concoction which is prepared by cooking bark from a special tree in water. It has the effect of causing the person to whom it is administered to vomit copiously. The conceptual connection between the witch vomiting and his getting rid of his witchcraft *shavi* is obvious. This is further emphasized by other details of the ritual: the witch is made to sit on an inverted pot positioned at some crossroads, while the *nganga* hits him with his whisk, breaking the pot under him; next he is led blindfold into a hut in his village which is then set alight; when he emerges from the burning hut he must run straight to his own house where he has to take a bath in water to which 'medicine' has been added. It is believed that after this treatment the person will no longer have dreams about becoming a witch, provided he has made an exhaustive confession of his sins at its beginning.

Vomiting also plays a role in the poison ordeals to which suspected witches used to be subjected in the past. If the 'medicine' caused the suspect to vomit, this was taken as proof of his innocence.[85] The same symbolism which is contained in the *nganga's* supposed ridding his clients of witchcraft by causing them to vomit is also contained in the emphasis which he places on their making a thorough confession about anything that could have made them vulnerable to the attack by the invisible powers.

Again, I stated above that the people's fear of witchcraft is so great that almost every village in the rural districts of Shonaland and many of the homes of Blacks living in urban areas have been protected by *nganga* against the entry of witches. This is done by driving 'medicated' wooden pegs into the ground and by placing 'medicines' under all footpaths leading into the homesteads.[86]

Finally, *nganga* are regularly asked by their clients to supply them with all kinds of protective 'medicines' and charms to preserve or enhance their 'luck'. As I have indicated, it is particularly in connection with their fulfilling these demands that they seem to trespass across the line which separates them from the witches, as the habitual wielders of power to do harm. Thus, for instance, *nganga* are said to supply men who have to be temporarily absent from their homes with a 'medicine' called *runyoka* which, when placed by them in their wives' beds, is believed to inflict a dreadful disease on any stranger who would dare to take advantage of their absence. Many similar 'medicines' are believed to be available from *nganga*, for the protection of a person's property, the recovery of stolen goods, as antidotes to witchcraft, and for numerous other purposes.[87] The problem is that not all of them serve such justifiable aims as does *runyoka* for example. And even in the case of *runyoka*, whoever attempts to take advantage of the home-owner's absence will regard as witchcraft what its user sees as a legitimate way of protecting his rights.

Similarly, a family may approach a *nganga* with the request to raise one of their deceased as an avenging spirit in order to let it exact vengeance against the people by whom they feel they have been wronged.[88] However, it will be remembered that witches are also said to be able to raise avenging spirits for their own evil purpose. While the raising of an avenging spirit is perfectly lawful in the eyes of the wronged party, the *nganga* who renders them this service is regarded as a witch by those who suffer from the spirit's vengeance. According to Crawford, even *chitsinga* and *chipotswa*, which have been seen to be typical witchcraft 'medicines', may be procured from a *nganga* for socially approved purposes. Therefore it is not surprising to find among the court cases he has recorded, one in which the accused were found to have tortured a *nganga* whom they believed to have traded in 'medicine for killing people'.[89] Similarly, Bourdillon tells of a *nganga* whose reputation fell considerably when she proved herself unable to kill off her husband's rival contender for succession to the chieftainship.[90]

All writers on the subject, therefore, stress the ambiguity of the *nganga's* position in Shona society, which accounts for the fact that he is treated with a mixture of respect, caution and fear.[91] The reason for this ambiguity is that 'Shona . . . regard doctor-diviners as partaking of the nature of a witch'.[92] As has been seen, there are obvious similarities between the witch and the *nganga* on both the conceptual and the practical level: the activities of the witch and the *nganga* are shrouded with secrecy,[93] they make extensive use both of 'medicine' and suggestive power to harm and to kill, and they are thought to be hosts to spirits who give them their ultra-human powers. Like the treatment of their clients, the *nganga's* power to divine the cause of death, illness, and misfortune is equally fraught with ambiguity.[94] Thus it will be remembered, that while a large part of his time is taken up by providing people with protection against witchcraft, it is through his divination that some people are said to be desired as hosts to witchcraft *shavi*. As the title of this chapter emphasizes, the rationale behind the belief in the powers of the *nganga* is the same as that behind the fear of witchcraft.

NOTES

1 Cf. Crawford (1967), pp. 58, 82-6; Gelfand (1959), pp. 122, 129, 150f.
2. Douglas (1970), p. XXIX.

3. Bourdillon (1972a), pp. 11, 51.
4. Bourdillon (1976), p. 201. Gelfand (1967), p. 26 writes that the deceased witch may pass her *uroyi* on to only one of her children or to all her offspring. According to him, '*uroyi* is the Shona word for both an anti-social act harming an individual and for the anti-social *animus* in the person carrying out the anti-social act': cf. *ibid.*, p. 25. For the present and the following paragraph compare the facts presented in Chapter Three, in the subsection *Maternal ancestral spirits.*
5. Aquina (1968), p. 50f.
6. Crawford (1967), p. 79. Cf. Gelfand (1959), p. 162; (1967), pp. 12, 20.
7. Lewis (1970), p. 293.
8. *Ibid.*, p. 303f.
9. Gelfand (1964), p. 45; (1967), p. 43. He uses 'black' in the sense it is used in the well-known term 'black magic'.
10. Bourdillon (1976), p. 205. My emphasis.
11. Cf. above, p. 98.
12. Cf. Crawford (1967), pp. 45-59; Gelfand (1967), pp. 46-8. Both authors stress the fact that due to the existence, since 1899, of the Witchcraft Suppression Act, relatively few cases arising from witchcraft imputation have come before the Rhodesian courts.
13. Crawford (1967), pp. 59-62.
14. *Ibid.*, p. 242f. Evidence given in court by the Member-in-Charge of the British South Africa Police post at Sipolilo. Mangava is described as an evangelist, and in this position he presumably held services in the school room on Sundays.
15. Bourdillon (1976), p. 213.
16. Douglas (1970), p. XXXV.
17. Murphree (1969), p. 185. His attitudinal survey among adults in a Budjga community revealed that the statement 'Witches exist' received an affirmative response from 90% of the Catholics and Methodists, 92% of the *Vapostori* (Members of an Independent Church), and 93% of 'non-Church people'.
18. Crawford (1967), p. 292. He writes that 'the Shona no more spends his ordinary waking hours thinking about wizardry than you or I spend our time thinking about the Atom Bomb'. Bourdillon (1976), p. 222 compares a Shona's continual awareness of the threat of witchcraft with the European's awareness that he is constantly exposed to germs.
19. Douglas (1970), p. XXXII.
20. Cf. also the analysis of accusers and victims in Crawford (1967), p. 133f and in his Appendix I.
21. Crawford (1967), p. 289.
22. Bourdillon (1967), p. 224.
23. Cf. Crawford (1967) and Gelfand (1967), *passim.*
24. Gelfand (1967), pp. 146-9. Cf. the remark, made above in chapter two, that the belief in the existence of a causal link between committing incest and harvesting a good crop is still popular among Shona people.
25. Hamutyinei and Plangger (1974), p. 280.
26. *Ibid.*, p. 281.

27. Kuper (1973), p. 628 says that in the eyes of the Swazi their king is potentially the greatest evil-doer, because they imagine him to be endued with the power of cosmic forces.
28. L. Makarius (1974) rightly stresses this aspect in an article entitled 'The Magic of Transgression', which on the whole, however, approaches the witchcraft phenomenon in a rather simplistic manner.
29. Cf. Gelfand (1959), p. 166f; (1967), pp. 27-9; Crawford (1967), pp. 123-5.
30. Bourdillon (1976), p. 211; also p. 227. Crawford (1967), p. 107 notes that the Shona word *chikambi* may mean both a witch and an anti-social person. Cf. also his p. 280.
31. Evidence presented in the following paragraph is based on Bourdillon (1972a), p. 269; (1976), pp. 210-214.
32. Bourdillon (1976), p. 214. Apparently the meaning of this statement is the same as that of a similar sentence on p. 202: 'Power is always associated with power to harm'.
33. Crawford (1967), p. 289. Cf. the similar criticism made by Douglas (1970), p. XVII-XX.
34. Aquina (1968) writes that *chitsinga*, a charm placed in the path of an enemy by male witches, causes the legs to swell up. They charge £2-£3 for returning their victim to health. Cf. also Crawford (1967), p. 267; Gelfand (1967), p. 41f.
35. Conversation overheard in 1900 in the kraal of a headman in rebel paramount Kunzi-Nyandoro's country: cf. Ranger (1970), p. 9. The Witchcraft Suppression Act had been passed during August 1899. The immigrating Whites were also regarded as 'magicians and wizards, unbeatable as fighters'. Cf. Young, (1970), p. 58.
36. Bourdillon (1976), p. 219.
37. 'When you marry, marry a witch; you will have somebody fighting on your behalf against a grudge', recommends a Shona proverb. Cf. Hamutyinei and Plangger (1974), p. 238.
38. Bourdillon (1976), p. 222.
39. Cf. Hemans (1970); Robinson (1971); Crawford (1967), pp. 197-202, 283.
40. For the sake of convenience, in the following I will use the shorter Shona term, *nganga*.
41. Bourdillon (1976), p. 192f.
42. Various types are in use and some *nganga* use all of these, throwing them in turn or in combinations. Cf. Gelfand (1964), p. 77f, and the photographs following pp. 72, 80; also Bourdillon (1976), pp. 176-9.
43. Cf. Gelfand (1967), p. 30: 'The first sign that she is destined to become a witch is the appearance of the spirit in her dreams . . . She dreams regularly that she is visiting graveyards and eating flesh from the bodies'.
44. Gelfand (1964), pp. 123-33; (1967), pp. 115-33.
45. Gelfand (1959), p. 121f.
46 Similarly, witchcraft *shavi* have often been seen to become 'naturalized' as ancestral spirits of particular families. Cf. beginning of this chapter.
47. Note the similarity between these symptoms and those indicative of a calling to become the medium of a territorial spirit. Weinrich, who has done research into the motives for which young African women have joined religious orders

in Zimbabwe, was told by one of them: 'If you have a vocation you are like a mad person; you just have to go ahead and become a Sister'. Another said she had dreamed as a child that she would become a Sister, and ever since she could think of nothing else: cf. Weinrich (1975), p. 229.

48. Cf. also Heintze (1967), p. 347 (map).
49. Cf. Gelfand's analysis of conditions treated by one particular *nganga:* (1964), pp. 169-77. Cf. also *ibid.*, pp. 134-43.
50. Cf. Crawford (1967), Appendix I. I have included prophets in this count, because they were found guilty of the same contravention of the Act as the *nganga*. In practice, their method of divining usually differs from that employed by the *nganga*. On the difficulty of keeping *nganga* and prophets apart, cf. *ibid.*, p. 204f.
51. Cf. Gelfand (1964), p. 88.
52. Crawford (1967), p. 190; cf. Bourdillon (1976), pp. 176-9.
53. Cf. Evans-Pritchard (1937), pp. 294-309.
54. Gelfand (1964), p. 76.
55. Crawford (1967), p. 192.
56. Bourdillon (1976), p. 179.
57. Cf. Gelfand (1964), p. 87; Bourdillon (1976), p. 176.
58. Cf., e.g. Crawford (1967), p. 196; Gelfand (1967), p. 145.
59. Crawford (1967), p. 195.
60. *Ibid.*, pp. 215-21, 283.
61. Bourdillon (1976), p. 181f. Cf. also Fry (1976), p. 35.
62. Crawford (1967), p. 193f.
63. Gelfand (1964), p. 88.
64. Bourdillon (1976), p. 175.
65. *Ibid., loc. cit.*; Crawford (1967), p. 191.
66. Gelfand (1964), p. 85.
67. Bourdillon (1976), p. 184f.
68. Cf. Evans-Prichard's (1937), p. 69f. classic exposition of Zande reasoning concerning the causation of misfortune by witchcraft. Bourdillon (1976), p. 199f. specifically refers to it when speaking of the same subject with regard to the Shona.
69. Cf. Gelfand (1964), p. 115.
70. Hamutyinei and Plangger (1974), pp. 125, 346. Cf. also pp. 230, 232.
71. Gelfand (1964), p. 115.
72. Bourdillon (1976), p. 172.
73. Gelfand (1973), p. 116. Robert Moffat, pioneer missionary among the Tswana and Amandebele, records Mzilikazi as saying to him when he took leave of him in 1857: 'How I love you! How can you say you would not come again? I shall like to reward you for your kindness . . . You gave medicine to the sick and you have cured my beloved wife. All the doctors in the land have been called. They could do nothing. I pay and feed doctors, but they are fools'. *Ibid.*, p. 114f.
74. Cf. Gelfand (1964), p. 71f; (1967), p. 44.
75. Crawford (1967), p. 108; Gelfand (1967), pp. 144-6.
76. Gelfand (1967), p. 43.

77. Bourdillon (1972a), pp. 75, 77.
78. Aquina (1968), p. 50.
79. She is often thought to use her familiars as carriers of 'medicines'. However, her 'medicines' are also believed to be able to seek out their intended victim by themselves, thus justifying the metaphor employed by Gelfand. It is commonly believed that witches are able to send a medicine-horn (*ngona*) to suck at the breasts of women, and thus to 'poison' their milk, causing their children to die: Crawford (1967), pp. 126, 241.
80. Crawford (1967), p. 51.
81. Cf. *ibid.*, p. 126; Gelfand (1967), p. 45. The latter writes (*loc. cit.*) that in medical terms, the conditions ascribed by the Shona to the action of these two 'medicines' can be described as polyarthritis, chronic rheumatism, lumbago, etc.
82. Gelfand (1967), p. 45f;Bourdillon (1976), p. 186.
83. For descriptions of various manners in which such exorcisms are attempted, cf. Crawford (1967), pp. 109, 271ff; Gelfand (1964), p. 47f; (1967), p. 108f.
84. Gelfand (1967), p. 109.
85. Crawford (1967), pp. 214-18.
86. Gelfand (1959), p. 168; (1964), pp. 50, 92; Garbett (1963), p. 53; Crawford (1967), p. 112.
87. For extensive lists, cf. Gelfand (1959), pp. 116-120; (1964), pp. 69-73; (1967), pp. 140-154. Concerning *runyoka*, cf. Gelfand (1964), p. 54.
88. Cf. Crawford (1967), p. 260f.
89. Crawford (1967), pp. 101, 178.
90. Bourdillon (1972a), p. 259.
91. Cf. Gelfand (1964), pp. 45, 120; Bourdillon (1976), p. 188.
92. Crawford (1967), p. 59. Bourdillon (1976), p. 206, essentially concurs with Crawford, qualifying, however, the latter's statement by saying that 'some people go so far as to say that all professional healers are witches'.
93. Cf. the proverbial expression: 'Diviners do not enjoy each other's presence'. Hamutyinei and Plangger (1974), p. 217.
94. 'The mouth of the *nganga* is a destroyer'. *Ibid.*, p. 266.

CHAPTER 6

THE SHONA 'CHURCHES OF THE SPIRIT': STORAGE TANKS FOR POWER

The Shona 'Churches of the Spirit' (*makereke dzoMweya*)[1] are the most common and most popular of the two major types of Independent Churches in Shonaland, the second type comprising what Daneel refers to as the 'Ethiopian-type Churches'.[2] Their origin dates back to the early 1920s, when individual Shona labour migrants became members of South African Independent Churches. These had grown partly out of contacts with similar movements in the United States. Their designation as 'Independent Churches' is justified by the fact that they are wholly administered by Blacks and most of them have also been founded by Blacks. It contrasts them with the Mission Churches in which the transfer of important positions of leadership from Whites to Blacks has been a slow process. Daneel points out that

> the frequent failure of Missions to cope with the real issues in the lives of traditionalists or of their African converts left a vacuum which the Independent Church leaders were well equipped to fill . . . Familiarity with existing practices and beliefs facilitates their task of presenting the tenets of Christianity to the African mind . . . in a more appropriate idiom.[3]

The 'Churches of the Spirit' accuse the Mission Churches most particularly of having 'repressed the message of the Spirit' and of 'withholding "Jordan Baptism" from Africans'. They claim that for this reason the Mission Churches lack the Holy Spirit. By contrast, the 'Churches of the Spirit' profess to have the 'real Holy Spirit referred to in the New Testament', through whom alone 'people can be helped to stop doing wrong'.[4]

The 'Churches of the Spirit' consist of two main branches, the largest of which is the African Apostolic Church of Johane Maranke (AACJM), whose members are commonly known as *Vapostori* (Apostles).[5] The other branch is made up of a congeries of movements known as Zionists, because each of them makes reference in its name to Zion, the favourite title given to Jerusalem in the Old Testament. The most important of these are the Zion Christian Church (ZCC), founded by Bishop Samuel Mutendi, and the *Ndaza*[6] Zionists. The latter fall into two major groups, Bishop David Masuka's Zion Apostolic Church of South Africa (ZAC of SA) and Bishop Andreas Shoko's Zionist Apostolic Faith Mission (ZAFM), from each of which have sprung a number of secessionist bodies.

There are no reliable figures available with regard to the numerical strength of the 'Churches of the Spirit'. Barrett puts the membership of all Independent Churches at eight per cent of the Black population in Zimbabwe,[7] which is nearly half as many people as are claimed to be members of the Mission Churches. However, Bourdillon warns that Barrett's figures are based on impressions.[8]

Only a decade after the Shona Rebellion of 1896, the first Independent Church was founded in Shonaland by Matthew Chigaga Zwimba. It deserves a brief mention, because some of its characteristics form a link between the Shona Uprisings and the 'Churches of the Spirit'.[9] His father had been made paramount chief of Zwimba Reserve by the White administration because of the loyalty he had shown to it during the Rebellion. Matthew became a catechist for the Methodists and established the first mission school in Zwimba Reserve. When, in 1907, he was transferred to another mission, outside his native chiefdom, he clashed with his White superiors and was dismissed from his office as catechist and teacher. In 1915, he founded the 'Original White Bird Mission', which declared as martyrs and saints the Zwimba men who had died during the Rebellion. The 'White Bird' of its title was a reference to both the common representation of the biblical Holy Spirit as a dove and to the mythological messenger of the territorial cult of Matonjeni, the 'little white bird of Mwari'.[10] He sent daring letters to Methodist superintendents and government officials, insisting that the Holy Spirit had commanded him to return to Zwimba, for 'Zwimba Mission Station was commenced by me . . . and I am going to commence [sic] the same work on it'.[11]

Zwimba's 'Original White Bird Mission' was primarily the reaction of a lone individual who had become articulate enough in the language of the foreign rulers to give expression to the mixture of defiance and despair which filled the Shona soon after their crushing defeat in the course of the Rebellion of 1896. It never gained a following beyond the Zwimba chiefdom and its activities were effectively checked by the White administration.

Although the founders of the first 'Churches of the Spirit' also encountered prolonged opposition – from the administration, the Mission Churches, and the tribal authorities[12] – their aim was different from the beginning. Few of them had ever been office-bearers in a Mission Church, and their foundations were not intended to counter the influence which the Mission Churches were beginning to gain over their fellow Shona. Rather they appear as men who were striving to interpret, in terms of familiar religious notions held by every traditional Shona, the new forces which had entered their country as a consequence of the preaching of the Biblical Message. I agree with Daneel's view that this direction of their work explains in large measure the popularity which their movements have gained among contemporary Shona. However, unlike him, I am not primarily concerned with discussing the degree to which this reinterpretation of the Biblical Message in familiar Shona terms may have

Distribution of Zion Christian Church congregations

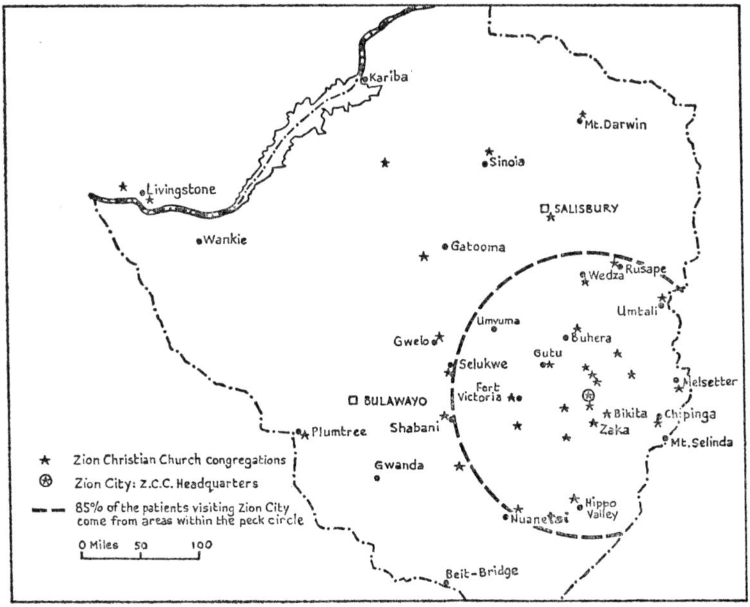

led to its falsification.[13] Instead, I wish to prove that the traditional conceptualization of spirits continues to be prominent in both the pronouncements and the actions of their founders and key officials.

Samuel Mutendi's Zion Christian Church

I am going to show that Samuel Mutendi's Zion Christian Church is largely conceived of as a counter-establishment to the territorial cult of Mwari of Matonjeni. However, before this I would like to present briefly some important biographical facts about the Church's founder.[14] Mutendi was born in Bikita in Southern Shonaland,[15] and was able to trace his descent through several royal houses to the Rozvi Mambos, Dyembewu and Chirisamuru. He received rudimentary training in reading and writing during a two year attendance at a Dutch Reformed Church (DRC) school, in which the main emphasis was placed on giving the pupils a basic knowledge about the Bible from a Protestant point of view. Mutendi became a full communicant-member of the DRC and served for some time as teacher-assistant in rural schools run by this Church. At one stage he also acted as a DRC preacher. Later he became a constable in the British South Africa Police.

When Mutendi went to the Transvaal in 1923, his principal motive for undertaking the long journey was 'to work in order to obtain cattle'. However, he claims to have had a dream in 1919 in which an angel told him: 'Look . . . you will have your own Church in this country'. While he and Andreas Shoko were working in Pretoria as bricklayers, they were attracted by the preaching of Enginasi Lekhanyane, the local leader of the Zion Apostolic Faith Mission. This Church had been founded by a certain Eduard in Lesotho and was one of several offshoots of the Zionist Apostolic Church, whose origins can be traced to the Christian Catholic Apostolic Church in Zion which was founded in 1896 by J. A. Dowie in the United States. Mutendi and Shoko were eventually baptized by Enginasi in 'Jordan' and shortly after his new conversion, Mutendi was chosen as 'Zionist emissary to Rhodesia' at a special Church meeting. After a year's campaigning in and around the Bikita district, Mutendi and other Rhodesian Zionist leaders were summoned to Pretoria because of an imminent schism between Lekhanyane and Eduard. However, only Mutendi attended the Zionist conference, during which Lekhanyane presented the constitution of what was to be known as the Zion Christian Church (ZCC). Mutendi was quick to join the new Church and was immediately ordained as one of its ministers. Equipped with the ZCC's credentials,[16] he returned to Rhodesia 'with more prestige than he had formerly had and with less binding obligations to the other Zionist leaders. He had, in effect, come back to build his own Church!'[17]

As I have described in chapter two, traditionalists regularly request rain, health, and fertility from the oracle at Matonjeni. Mutendi, for his part, concentrated on satisfying precisely these widely felt needs.[18] His followers are not permitted to participate in the annual rain rituals (*mukwerere*) at Matonjeni, which are held there in honour of Mwari at the beginning of the rainy season in October. They are also dissuaded by their leaders from giving gifts to Mwari's messengers which these collect for presentation to the oracle. Instead, each Zionist congregation sent a delegation with its special request for rain to Mutendi's 'Zion City' at Gutu,[19] two hundred and fifty-six kilometres east of Matonjeni, for the October 'seed conference'. During it, Mutendi would bless the seed which was soon to be sown, as well as the pegs which ZCC members positioned at the corners of their gardens, to protect their crop against witchcraft. Surrounded by hundreds of his followers, the founder of the Church would preside over the dancing of the delegations. These came up to him one by one, each of them wanting to present him with the largest gift of money, in recognition of their faith in his God-given power to provide their district with an abundant share of rain.[20]

Zionists have also substituted an alternative ritual for the traditional Shona rain-petitioning and seed-blessing ceremonies which are held each year at the land shrines of territorial spirits. They organize large-scale fasts on the tops of mountains, lasting up to three days, in which Church members from several neighbourhoods join together. Through their action they endeavour to 'clear the house', lest God retaliate with a bad rainy season or with illness. On its arrival, each group is scrutinized by prophets, who line themselves up in pairs, to form the 'gate' to the mountain. Passing the 'gate', each participant must make a public confession of his sins. The prophets prevent from passing, whomever the Holy Spirit indicates to them has failed to make an exhaustive confession. This penetrating 'clearing of the house' continues on top of the mountain, helped along by the prophets through outright accusations or subtle suggestions, made in the name of the Holy Spirit.

ZCC members would reject outright the suggestion that to them the Holy Spirit is a new kind of territorial spirit who uses Mutendi and the prophets as his mediums. Nevertheless, it is clear that in their minds much of the traditional Shona thinking about territorial spirits has become part of their notions concerning the Holy Spirit.[21] For this reason, in their sermons, Mutendi and his subordinate leaders would refer constantly to Mwari of Matonjeni as Satan and claim that this traditional territorial spirit no longer has the power to provide people with rain. Together with the territorial spirits who were the traditional 'owners of the land', they denounce the ancestral spirits, the traditional 'guardians' of the family, as evildoers. Preachers of 'Churches of the Spirit' encourage their followers to 'cast away all these things and [to] believe in Him who is in heaven'.[22] It may seem rash to liken Mutendi to a spirit medium on the scant evidence I have presented so far, but the similarity will become more obvious.

One indication as to the correctness of my interpretation is the reaction which Mutendi's activities have elicited from the chiefs and spirit mediums, i.e. the two traditional tribal authorities. Many of the former were won over by Mutendi as followers after he had convinced them that he could 'make rain' almost at will. Some of them had formerly been his strongest opponents. As regards the spirit mediums, their sharp reaction to the rain-making rituals of Mutendi and other leaders of 'Churches of the Spirit' indicates that they regard these as effective replacements of those rites on which their own influence over the people is based. Traditionalists, therefore, often accuse members of these Churches of 'polluting the earth', because their refusal to participate in the traditional rites is believed to anger the territorial spirits, causing them to withhold the rain or to destroy the crops. Naturally, Zionists and Apostles counter this accusation by blaming droughts on the unwillingness of the former to turn to the Christian Mwari.[23]

Mutendi's declaration that the territorial and ancestral spirits are Satan and demons may be more than a crude attempt to undermine the esteem these have in the eyes of Shona traditionalists. I pointed out in chapter two that, the founder-ancestors of Shona chiefdoms are invariably portrayed as people who were not afraid of going against generally accepted norms, and that by doing so they were believed to have acquired prodigious powers over man and nature. One would expect that daring to refer to the traditional 'owners of the land' and 'guardians' of families as evil spirits would endow the founders of the 'Churches of the Spirit' with a similar aura of invincible power. The fact is that in both branches of these Churches the two most important ones possess a Holy Book which records the exploits of their founders and glorifies their miraculous powers.

The ZCC's Holy Book, entitled *Rungano*, narrates the growth of Mutendi's Church in spite of opposition from the White administration, tribal authorities, and missionaries of the Dutch Reformed Church.[24] It also tells how Mutendi's awe-inspiring powers once prevented a number of discontented leaders from seceding from his Church. During a crucial meeting with them, he is reported to have said, 'If God is with me I will have Him send down lightning so that you will be thrown to the ground because you do not listen!' The rain and lightning came directly after the meeting. It was of such force that one of the discontented leaders ran to Mutendi's house, begging the 'Man of God' to come to their aid. Mutendi then pointed his staff at the clouds and with a loud 'Amen!' caused the storm to subside. All were filled with great fear and went to Mutendi to confess the evil thoughts they had harboured about him.[25]

Johane Maranke and the African Apostolic Church

Johane Maranke, the founder of the African Apostolic Church, was born in 1912, in the Maranke Tribal Area[26] and, like Mutendi, was of royal blood. His mother was a daughter of Chief Maranke and had become the medium of one of the senior territorial spirits. He attended a Methodist mission school and was baptized, being given the name Roston. After completing his fourth year in school, he worked as a labourer in Umtali. He is said to have had visions from the time he was a boy. However, his decisive calling by the Holy Spirit occurred in 1932 near Mt. Nyengwe, in his home district. He suddenly became aware of a bright light shining down on him and he heard a voice saying: 'You are John (Johane) the Baptist, an Apostle. Now go and do my work'.[27]

The African Apostolic Church's Holy Book, the 'New Revelation of the Apostles' (*Umboo utsva hwavaPostori*), is regarded by its members as a canonical addition to the Bible. This is how the book describes the conferring of extraordinary powers on the Church's founder by the Holy Spirit:

> You will have power to cure the sick by laying hands on them and by consecrating water for them to drink; you will be able to drive away any kind of *shavi* through the laying-on of hands. Take the long staff, wherever you go, for healing purposes! Through your hands fertility will be conveyed to the barren and when you step on fire you will not be burnt . . . When you want to lay hands on a new convert so that he can perform miracles, so be it.[28]

Importance of Mutendi the founder

The importance of the founder is stressed throughout the *Rungano* a d the *Umboo utsva hwavaPostori*, as well as in the ZCC's constitution. Under the heading 'Executive', the latter declares Samuel Mutendi as 'Life Senior Elder Founder' and 'Director Superintendent', and decrees that upon his death one of his twenty-three sons is to be elected to hold the position of 'Senior Eldership'.[29] As founder of the ZCC, Mutendi is declared by the *Rungano* as 'owner of the job', who has full control over it and, according to Daneel, he is to his followers 'in a certain sense the Church itself'.[30]

Both Holy Books also abound with instances which are obviously intended to prove that the founders of these two Churches are equal in power to Jesus Christ and other important persons of the Old and New Testaments. The *Rungano* even contains an example of the kind of manipulation of myths which have been seen to be employed by traditional spirit mediums and chiefs when they are making rival claims to the ownership of land.[31] It likens Mutendi and the cord-wearing (*Ndaza*) Zionist Bishop David Masuka to Moses and Aaron implying that compared to Mutendi Masuka holds an inferior position.[32] It is also significant that all founders of 'Churches of the Spirit' have expressed their desire to associate themselves with the powerful beginnings of Christianity through the names which they have chosen for their foundations. These refer therefore, to Zion, the Apostles, the Holy Spirit, etc.[33]

I have mentioned the opposition which Mutendi in particular encountered in the early years of his preaching activity in Rhodesia.

Chiefs accused him of 'spoiling the country', alleging that he 'plunged their children into the rivers' – a reference to his 'Jordan Baptism'[34] and that his preaching and prophecies aroused maneless lions which attacked their cattle at night.[35] Between 1923 and 1932 he was, therefore, repeatedly retained by the authorities and forbidden to preach. However, whereas during this period the power of chiefs was rapidly declining, the effective control of tribal affairs being securely in the hands of Native Commissioners, the *Rungano* suggests that its dauntless hero won even the respect of the Commissioners by performing astounding miracles while he was in their custody. Of the chiefs, on the other hand, the Holy Book tells that they had to suffer the humiliation of being called 'bloody fools' by the Native Commissioners.[36]

The subsequent change of attitude towards Mutendi among the tribal rulers is reflected in the fact that in 1965 his following included fifteen chiefs and sub-chiefs, the largest number of notables any Independent Church in the country was able to claim.[37] His 'Zion City' is a bustling township of well over 200 huts, and larger than any other headman's village. It is the centre of Mutendi's widespread sphere of influence, at which every year hundreds of his followers congregate for the performance of rain-making ceremonies and other rituals. As a kind of latter-day Zimbabwe, it 'appealed to the sentiment of tribal leaders (as much as it aroused the envy of others), who had not forgotten the glories of the once-powerful Rozvi dynasty'.[38]

There is also evidence that an inter-dependence is developing between leaders of Independent Churches and the chiefs, similar to the traditional relationships between spirit mediums and chiefs. Mutendi arbitrated in border disputes between chiefs, and chief Munyonga, whose territory had always been closely associated with the Matonjeni oracle, relied on his advice during a difficult succession dispute, unconcerned about the oracle's threat to retaliate by causing a drought. Leaders of Independent Churches, on the other hand, appeal to their local chief's court when they are in trouble, particularly during crises of leadership.[39]

Struggles for leadership play a large role in the Independent Churches and various sample surveys carried out by Daneel have revealed that eighteen per cent of the people defecting from these Churches were motivated by dissatisfaction with Church leaders and office-bearers.[40] A major crisis usually develops after the death of the founder, but the reason for this is not as obvious as it would seem at first sight. It was noted in chapter two that due to the complicated rules of succession it often takes considerable time before a successor can be appointed after the death of a Shona chief. However, one would not expect this problem to arise at the death of a Church founder. According to traditional rules, chiefly office should be inherited by the eldest son. As I have mentioned, in contrast with this traditional rule, the constitution of the ZCC stipulates that upon the death of Mutendi 'the *Church* will choose one [of his 23 sons] with exemplary character to be the Bishop in place of his father'.[41]

The constitution fails to make it clear who shall form part of this elective body. Presumably it will consist of all those who are entitled to attend the Church's Annual Conference, i.e. twelve nominally mentioned ordained ministers; an unspecified number of lay preachers;[42] and two delegates from each congregation of no less than 25 persons, to be elected from among its Church officers. At the ZCC's present membership of some ten thousand followers, spread over the whole of Zimbabwe, there would thus be a large number of delegates involved in the election of Mutendi's successor, in addition to the ministers and lay preachers. Each of them is a potential medium through whom the Holy Spirit may indicate his preference for one of the twenty-three candidates, or indeed reject all of them, favouring instead a powerful personality among the electorate themselves.

Succession Dispute after Maranke's death

The succession dispute which erupted in 1963 after the death of the African Apostolic Church's founder, Johane Maranke, is a case in point. It has led to that Church's biggest schism. For this reason it deserves a brief description. However, it is necessary first to sketch its background.

The night after Johane had his above-mentioned vision, in which the Holy Spirit conferred on him his special powers, his relatives saw him seized by continuing spells of spirit possession. In the morning, Conorio and Anrod, his elder siblings, and their paternal uncle, Peter, went out and told people in the neigbourhood about the strange events of the previous night. Significantly, however, Johane first of all went to tell Ruka Mataruka, his wife's sister's husband, of his vision of a new Church. Ruka became his first official convert. Johane caused him to be filled with the Holy Spirit and made him his 'Top Evangelist'. It is noteworthy that a number of further important posts in the new Church went to relatives of Johane's who did not belong to his agnatic line. Thus Simon Mushati, his mother's sister's son, became his 'Top Prophet',[43] and another maternal cousin was made the Church's secretary. Although his father and two brothers soon became his followers, during the first historic baptism the Holy Spirit caused Ruka Mataruka to 'fall into the pool' before everyone else. Ruka baptized Johane, who in turn first baptized Simon, then Ruka and a further 150 converts.

Anrod, Johane's elder brother, became a junior prophet and as such was subject to Simon's authority. Conorio, his eldest brother, was made 'Top Healer', while Johane's father became a 'Judge' in the new Church. When Conorio died without leaving sons to continue his work, his position was inherited by Anrod. However, compared with him, Simon Mushati was far more prominent in the Church, accompanying Johane on most of his extended tours, and being widely regarded as his second-in-command. When Johane fell gravely ill during one of these tours and soon afterwards died, Simon and another Church official who had accompanied the dead leader on his last tour, were accused of having caused his death through witchcraft in order to succeed to his leadership.[44]

It is reasonable to assume that Johane conferred the most important leadership positions in his Church on persons outside his agnatic line to prevent his senior brothers and other agnatic relations from challenging his supreme leadership. However, upon his death, the agnatic lineage asserted its right to inherit the deceased's authority and property. As Johane's brother, Anrod invoked the traditional Shona law of succession, on behalf of Abero, the founder's senior son. As 'father's eldest brother', he presided over the inheritance ceremony which was held in the traditional manner, without the use of beer, on the day after Johane's burial.[45] During it, Abero received Johane's hereditary name and his staff, thereby accepting responsibility for his father's family, and his installation as 'First Priest' of the African Apostolic Church. Following this, Anrod handed

Johane's other two staves to Makebo and Judah, Abero's younger brothers, conferring on them the second and third leading positions in the Church.

At this juncture, Simon stood up and complained to the meeting, stating that 'the inheritance of the Word of God is not inherited by children'.[46] However, he was silenced by Anrod, who pointed out that as a distant relative Simon had no say in the matter. He also claimed to have privately received Johane's last instructions. According to these, Simon was to 'rest' and cease going round on tours to the Church's outlying congregations. As mentioned above, his enemies had already given Simon the reputation of being a witch. Now they spread in addition the rumour that he had been planning all along to organize his own Church. After further squabbles and an unsuccessful attempt on his part to get satisfaction by an appeal to the local chief's court, Simon Mushati did in fact break away from the Church which had been founded by his cousin, and formed his own 'African Apostolic Church St. Simon and St. Johane'.[47]

This brief sketch of its history reveals a number of noteworthy features about Johane's African Apostolic Church, but two of these should be singled out because of their importance: firstly, the strong emphasis this Church, like other Independent Shona Churches, places on coupling rank with authority; and, secondly, the use of traditional rules of inheritance to determine succession in all categories of Church leadership. In view of the failure of the Zion Christian Church's constitution to regulate Mutendi's succession in a precise manner, it is not surprising that these two features manifested themselves following the Church founder's recent death. As Daneel notes,[48] Mutendi's followers decided to wait a full year before they would attempt to settle the leadership issue. Of his sons, Ruben was the only Church minister prior to Mutendi's death. He is confident that he will succeed his father. However, many ZCC members prefer Enginasi, another of Mutendi's sons, who played an important part in Church Council affairs when his father was still alive. To complicate matters, there is Jainos, the most senior son of Mutendi, who was, however, an inactive Church member. Daneel believes that Jainos will probably take over the headmanship of 'Zion City' and says that 'it is just possible that he will stake a claim to Church leadership, if he can get the support of those regional leaders who do not want Ruben or Enginasi'. He also notes that there are bound to be claims and counter-claims based on prophetic revelation.

Secessions

However, even before the death of its leader, the ZCC had already produced no less than thirteen splinter groups between 1929 and 1961, whose founders seceded on account of some major disagreement.[49] Daneel ends his description of the succession dispute within the African Apostolic Church with the remark: 'Anrod was prepared to face a schism rather than allow Simon to assume power within the Church'.[50] It can be safely held that in like manner the majority of Independent Church leaders in Shonaland accept schisms as the only solution to a complex problem of leadership. Very often secession is triggered by a successful campaigner's wish to assert full authority over the church members whom he has newly won in a particular area.[51] It will be remembered that this was the reason for Zwimba founding his own 'Original White Bird Mission'. Often too, these secessions are preceded by an uneasy period of tension, during which officials of the two parties preach 'fighting sermons' against each other during Church services. In one of these, for example, Anrod called Simon a heathen, because he allegedly attempted to cause a rift in the Church. Simon countered the attack by claiming that Anrod had illegally grabbed the power in the Church, for, he said: 'Is it not I who had helped Johane by laying hands on new leaders when they were consecrated?'.[52] It is quite obvious that centrifugal forces similar to those which frequently lead to the breaking-up of Korekore hamlets are at work here.[53]

Questions of rank and seniority are another cause of secessions. Seniority was at the centre of the dispute between Anrod and Simon. The latter was able to claim to have been the first person to be baptized by Johane[54] and to have functioned as second-in-command to the founder. Similarly, as founder of his own Zion Christian Church in Rhodesia, Samuel Mutendi no longer felt subject to the authority of Eduard Lekhanyane, the head of the Zion Christian Church in the Transvaal. It was Eduard's father, Enginasi Lekhanyane, who had sent Mutendi as 'Zionist emissary' to Rhodesia to found the Zion Christian Church there. As Eduard's senior, Mutendi regarded Enginasi's successor as 'a "young one" (*muduku*) whom he respected as the son of a friend, but who had no real authority over the Church in Rhodesia'.[55]

Naturally, the availability of the Bible as a written standard, against which one can measure the discipline and orthodoxy of leaders, is another weapon in the hands of pretending secessionists. Cases quoted by Daneel involve the question of keeping the Sabbath and reaction to the lax application of Church laws concerning marriage and beer drinking. However, he points out that often 'a doctrinal conflict only serves as a camouflage for the ambition of Church leadership, . . . the most potent source of separation'.[56]

As in the hamlets described by Garbett, where each of the senior males desires to eventually found his own hamlet, there is a 'continual shuffling of power' in Shona Independent Churches.[57] Occasionally it even happens that a Zionist undergoes the protracted Catholic catechumenate and has himself baptized a Catholic, only to quit attending Catholic services immediately following his baptism. Afterwards, such men set themselves up as Zionist leaders who believe themselves to be more powerful than before, because they 'think that Catholic priests are more powerfully possessed by the Holy Spirit than any other person'.[58] However, many secessionists fail to achieve their aim of building up a lasting, if not expanding, group of followers. Significantly, a number of them have subsequently returned to traditional forms of worship and one of them even ended by establishing himself as a *nganga*.[59]

Founders of Independent Churches and the Shona high god

More than ten years ago, Robin Horton caused a flurry among anthropologists with his theory that primitives, like scientists, were constructing theoretical models. Like the theoretical models of scientists, Horton claimed, primitive belief systems were also intended to explain and predict, answering such basic questions in life as 'Why did it happen?' and 'What can we do about it?'.[60] In a further elaboration of his claim, Horton insisted that in the case of both the scientist and the primitive, 'the quest for explanatory theory . . . involves the elaboration of a scheme of entities or forces operating "behind" or "within" the world of common-sense observations'.[61]

In 1971, Horton expanded his 'Intellectualist Theory' by adding an important point which is still under discussion in anthropological circles and which has a direct relevance to the subject of this chapter.[62] Put as briefly as possible,[63] Horton claimed that a 'basic' African cosmology has a two-tier structure, the first tier being that of the 'lesser spirits' and the second that of the supreme being. The 'lesser spirits' are the 'underpinners' of events and processes in the 'microcosm' of the local community and its environment, whilst the supreme being is the 'underpinner' of events and processes in the 'macrocosm', i.e. 'the world as a whole'. Just as the various microcosms form part of the macrocosm, so the 'lesser spirits' are conceived either as manifestations of the supreme being or as entities ultimately deriving their power from him.

In the pre-modern situation – marked by subsistence farming and poorly developed trade links – most events affecting the life of the individual occur within the microcosm of the local community which is considerably insulated from the wider world. This sort of situation, argues Horton, is likely to favour a religious system in which a great deal of attention is paid to the 'lesser spirits' whilst very little attention is paid to the supreme being.

However, as soon as this situation is changed through factors making for wider communication – e.g. the development of long-distance trade – social relationships will increasingly cut across the boundaries of the microcosms. As a consequence, Horton further claims, the religious forms will also change in that society: the 'lesser spirits' will now receive less attention, while the supreme being will receive more.

Against the background of the scheme as it has just been outlined, Horton then carries out a 'thought-experiment'. He confronts this 'basic' African cosmology with an imaginary set of changes and uses his model of its working to predict its response to these changes. He visualizes a massive development of commerce and nation states in Africa, without the influx of Islamic and Christian 'proselytizers' which has in fact been a concomitant of Africa's colonization. In this imaginary situation, Horton predicts, microcosmic boundaries will weaken and many people will be 'consequently groping toward a more elaborate definition of the supreme being and a more developed cult of this being'.[64]

In Horton's view, his 'thought-experiment' reduces Islam and Christianity to the role of catalysts. They only stimulated and accelerated changes which were 'in the air' anyway. He believes it further provides a theory that accounts for the failure as well as for the successes of the two world religions, because even where a rapid breakdown of boundaries between microcosms occurs,

> acceptance of Islamic or Christian patterns will be highly selective. Just what is accepted and what rejected will be largely determined by the structure of the 'basic' cosmology, and by the limits which this structure imposes on the cosmology's potential for response to social change.[65]

How useful is Horton's Intellectualist Theory of African Conversion for explaining the emergence of the 'Churches of the Spirit' among the Shona? The salient points of the theory in relation to the Shona are as follows:

1. Christianity is only a catalyst.
2. The acceptance of Christian patterns is highly selective. It is determined by:
   a. The structure of the 'basic' (Black) cosmology
   b. The limits which this structure imposes on the cosmology's potential for response to social change.

Considering first of all Horton's point 2), it is obvious that the acceptance of Christian patterns on the part of the 'Churches of the Spirit' has been carried out in a highly selective manner. This has been so, according to Horton, because the process of selection is determined by the structure of the people's 'basic' cosmology. In considering the cosmology of the Shona, it is important to note that the idea of a high god was not common among them in pre-modern times. As I have shown, the nearest equivalent to the notion of a high god has only arisen where the cult of a particular senior territorial spirit, like that of Mwari, Nehanda, Chaminuka, or Kaguru, has spread its influence over a wide area.[66] Still, it should be remembered that in Horton's scheme territorial spirits definitely belong to the 'lesser spirits', and also, that their influence has always depended upon the reputation which their mediums, messengers or priests have held among the people.

This fact places in question the value of Horton's scheme for predicting the reaction of the Shona's cosmology to the changes which have taken place in their country since the late nineteenth century. According to this scheme, as the various microcosms of a society move together more closely due to the changes brought about by colonization, the influence of the 'lesser spirits' among the people is said to decrease and that of the high god is supposed to increase. The following is what in fact did happen in Shonaland:

1. While it is true that today Mwari is widely known among the Shona – even among those adhering to the traditional religion – as a more or less remote high god; nevertheless, the reason for this is quite different from that stated in Horton's prediction. According to him,

> if thousands of people find themselves outside the microcosms, and if even those left inside see boundaries weakening if not actually dissolving, they can only interpret these changes by assuming that the lesser spirits (underpinners of the microcosms) are in retreat, and that the supreme being (underpinner of the macrocosm) is taking over direct control of the everyday world. Hence, they come to regard the lesser spirits as irrelevant or downright evil. Hence too, they develop a far more elaborate theory of the supreme being and his ways of working in the world.[67]

In fact, as has been seen, the widespread recognition of Mwari as high god derives from the adoption of this name by the Mission Churches in translating the name of the Biblical God into the Shona language. Its constant use in the Churches and in Bible classes in the schools for a considerable length of time is the foremost reason for the fact that today it is rather generally known.

2. Long before the teachings of Christian 'proselytizers' had reached significant numbers of Shona, and soon after the occupation of large parts of their country by the Whites, a small number of influential mediums of territorial spirits – who are obviously 'lesser spirits' – fomented and organized the Rebellion of 1896. As I have pointed out, there was a tendency during and after the Rebellion to emphasize the ultra-human powers of these territorial spirits, as well as their supra-tribal authority. However, this tendency lay within the predictable capacity for development of the territorial spirit/founder-ancestor concept, and is a far cry from Horton's postulated reasons for the growth of notions about a high god in modern circumstances.[68]

3. At the close of the thirty-odd years of apathy and depression following the bloodbath which ended the Rebellion – and after a period of concerted attacks against the Shona's traditional religious system by the Missions, mission schools, and the administration – Shonaland was swept by a veritable epidemic of alleged possessions

of people by stranger-spirits. Again there can be no doubt that these too were 'lesser spirits' and that their massive appearance was directly connected with the onslaught of the modern situation upon the traditional Shona microcosms.[69]

It is important to note that the beginnings of the 'Churches of the Spirit' fall into the same period. As will be seen in the next chapter, their founders were in fact accused of recruiting followers by inducing in them a state of mass hysteria similar to that which was thought to be typical of traditional possession seances involving stranger-spirits.

4. It is also clear from the evidence in the next two chapters that the emergence of the 'Churches of the Spirit' has not brought about a decrease in the importance of the traditional 'lesser spirits' of the Shona. It is true that territorial, ancestral, and stranger-spirits are denounced in these Churches as demons and Satan; however, the 'high god' in whose name they are thus denounced and exorcised, fits very well into the traditional Shona concepts of power and powers.

The Holy Spirit in the Shona 'Churches of the Spirit' is not the high god which should result, according to Horton's Intellectualist Theory, from the impact of modernity upon a 'basic' Black cosmology. Rather, the founders of these Churches tend to be regarded by their followers as quasi-divine, because they see them as the powerful mediums of a Holy Spirit, whose immanence in their leaders is largely conceptualized along the lines of a traditional territorial spirit's 'coming out' in his mediums.

5. One should also point out that the founders of the 'Churches of the Spirit' – and, to a lesser degree, their subordinate office-bearers – have accumulated in their persons nearly all the various sources of power which in the traditional Shona cosmology were attributed to different persons or spirits:

Like chiefs, they glory in the ownership of representative centres of power and enjoy the support of a large group of followers; like senior mediums of territorial spirits they function as rainmakers, guarantors of fertility, mediators in disputes between chiefs, and they are believed to be able to bring down severe punishment on any deviator; one's membership of the community they have founded ensures continued health and blessings, and a place in heaven – benefits which were formerly safeguarded by the ancestral spirits. Just as the hosts of powerful stranger-spirits are believed to be able to 'send' these spirits into other people, the founders of 'Churches of the Spirit' pass on the Holy Spirit through the laying-on of hands; and like traditional diviner-healers, they diagnose the cause of their clients' troubles, dispense 'medicines' intended to heal and to protect, detect witches and attempt their cure.

In conclusion, it can be said that Horton's Intellectualist Theory has failed to correctly predict the manner in which the Shona cosmology has reacted to the profound changes which have taken place in Shonaland over the past ninety years. For this reason also the Theory's ascription to Christianity of a mere catalyst role remains, as far as the Shona are concerned, an unverified hypothesis.[70]

NOTES

1. Cf. Daneel (1971), pp. 285, 347.
2. Cf. *ibid.*, pp. 350-385. To match this term, Daneel refers to the first category as 'Spirit-type Churches': cf. *ibid.*, pp. 286-349. In contrast to the 'classical' Ethiopianism of South Africa, described by Sundkler (1961), p. 53ff, Ethiopian Churches in Rhodesia are not militantly anti-White.
3. *Ibid.*, p 451.
4. Daneel (1974), p. 27f.
5. It was first described by Aquina (1967). Cf. also Murphree (1969).
6. Shona word for 'holy cord'. An early description of this group is to be found in Aquina (1969).
7. Barrett (1968), p. 78. According to him, in 1967, the figures for some of these Churches were as follows: AACJM – 10 000 members and 50 000 adherents in Rhodesia, 40 000 in Zaire. ZCC – 12 000 members. ZAC of SA – 12 000 adherents. Cf. *ibid.*, pp. 292, 296f, 303.
8. Bourdillon (1976), p. 332.
9. For the following, cf. Ranger (1970), pp. 20-5.
10. Cf. above, chapter two, subsections: *Territorial spirits and spirit cults*, and *The Shona Rebellion.*
11. Ranger (1970), p. 21. Interestingly, Matthew regarded the Zimbabwe ruins as proof that there had been a previous invasion of Shonaland by the Whites. He said the latter were now again building large towns all over Rhodesia, but the Natives would drive them out of the country as they had done before. Cf. *ibid.*, p. 22.
12. Peaden (1970), p. 65 reports that between 1924 and 1933 the '[Rhodesian] authorities were torn between the need for religious toleration and the need to control independent tendencies' among the Blacks. As will be seen in the next chapter, conflict arose particularly over the illegal opening of schools and the

conducting of religious services in remote places at night-time, by the Zionist preachers.
13. Cf. Daneel (1971), p. 3f and (1974), p. 309f.
14. For the following, cf. Daneel (1971), pp. 287-93.
15. Bikita is roughly ninety-six kilometres east of Ft. Victoria.
16. According to Peaden (1970), p. 55 his new credentials gave his name as Samuel Makuwa. As Peaden shows, the changing of their names was common practice among Zionists of the day and increased the state of confusion and suspicion in which the authorities found themselves.
17. Daneel (1971), p. 299.
18. For the following, cf. Daneel (1971), pp. 85, 89, 330f, 432f, 436 and (1974), pp. 103-13.
19. Details about 'Zion City' will be presented later in this chapter and particularly in chapter eight.
20. The African Apostolic Church, founded by Johane Maranke, holds similar annual gatherings for its members. These are, however, held at *Paseka* (from 'Pasch') or *Pendi* (from 'Pentecost') centres in each Tribal Trust Land, to which the founder or his close collaborators went in turn. By 1940 the Church had already well over 100 such sites and in 1965 more than 10 000 Apostles were reported to have received the sacraments during some of the *Paseka* ceremonies. Cf. Daneel (1971), p. 330f.
21. In his description of the Apostolic Sabbath Church of God, another Rhodesian 'Church of the Spirit', Bourdillon reaches the same conclusion. Cf. Bourdillon (1977), p. 208.
22. Daneel (1974), p. 104.
23. Cf. Garbett (1963), pp. 187, 196f; Daneel (1971), p. 438.
24. Cf. Daneel (1971), Appendix IV (Excerpts). The Shona word *Rungano* means 'history'.
25. Daneel (1971), p. 314.
26. Roughly eighty kilometres south-west of Umtali.
27. Daneel (1971), p. 319. Further details concerning the life of Johane Maranke will be given as required in the course of the argument.
28. Daneel (1971), p. 320.
29. *Ibid.*, p. 495.
30. *Ibid.*, p. 491, 310.
31. Cf. above, chapter two.
32. Daneel (1971), p. 426f.
33. This observation appears in Daneel (1971), p. 344f, though with a somewhat different slant.
34. *Ibid.*, p. 408. A detailed description of 'Jordan Baptism' will be presented in the following chapters.
35. *Ibid.*, p. 483. Note the specific association of these lions with territorial spirits. Cf. above, chapter two.
36. Cf. *ibid.*, pp. 408-411 & 484f.
37. *Ibid.*, p. 437.
38. *Ibid.*, loc. cit.,; cf. also Daneel (1974), p. 204
39. Cf. Daneel (1971), pp. 337, 437; (1974), p. 109f.

40. Daneel (1974), p. 81.
41. Daneel (1971), p. 495. My emphasis.
42. The constitution says 'all lay preachers'. Cf. *ibid.*, p. 496. Next in rank to the ordained ministers are eleven mentioned Evangelists, but they are not said to form part of the Annual Conference, unless they too are to be regarded as 'ordained ministers'.
43. Daneel (1971), p. 330 writes that in 1940 increasing trouble with leadership conflicts on the level of local congregations forced Johane to introduce a differentiated leadership structure in his Church. It consisted of a hierarchy of five different offices of leadership, starting with that of Baptist, and followed by the offices of Evangelist, Healer, Prophet, and Judge. Each office was subdivided into five ranks. Promotion from a lower to a higher rank depended on the death or disqualification of a senior, or on inheritance.
44. Cf. Daneel (1971), pp. 316; 321-4; 328.
45. Among the Manyika, to which Shona group Johane belonged, the 'bringing-home' (*bona*) ritual is traditionally held immediately after the burial. Apostles are not allowed to drink beer.
46. Daneel (1971), p. 334.
47. *Ibid.*, pp. 333-8.
48. Personal communication dated 10.3.77.
49. Cf. the 'Zionist genealogical table' and Bishop Mutendi's diagrammatic representation of Christianity in Daneel (1971), pp. 300, 493. All other Independent Churches treated by Daneel have suffered from similar secessions, but the ZCC has had the largest number of them. Cf. *op. cit.*, pp. 302, 331, 365, 374.
50. Daneel (1971), p. 338.
51. Cf. Daneel (1971), pp. 304, 331f, 339.
52. *Ibid.*, (1971), p. 336.
53. Cf. above, chapter three.
54. Ruka Mataruka, who, it will be remembered, baptized Johane, the founder of the Church, later challenged the latter's leadership over all Apostles. However, he lost an ordeal during which he allegedly 'made fire on his head' while preaching. Johane outdid this by placing his Bible and garments in a fire during his sermon and retrieving them from the fire unharmed. Cf. Daneel (1971), p. 332.
55. Cf. *ibid.*, pp. 291, 293, 299. On p. 305 Daneel records a case in which the son of a Church founder refused to be ordained by bishops who had been his father's subordinates.
56. Daneel (1971), pp. 302f, 309.
57. *Ibid.*, p. 306.
58. Aquina (1969), p. 133.
59. Cf. Daneel (1971), pp. 310f, 332.
60. Cf. Horton (1962), p. 208.
61. Horton (1967), p. 51.
62. Horton (1971).
63. *Ibid.*, pp. 101-3. I partly follow here Horton's own summary of his revised Intellectualist Theory in Horton (1975), p. 219f.

64. Horton (1975), p. 220.
65. *Ibid.*, *Loc. cit.*
66. Cf. above, chapter two.
67. Horton (1971), p. 102.
68. Besides the one already quoted, his second reason is the following (Horton [1971], p. 102): 'As more and more people become involved in social life beyond the confines of their various microcosms, they begin to evolve a moral code for the governance of this wider life. Since the supreme being is already defined as the arbiter of everything that transcends the boundaries of the microcosms, he is seen as underpinning this universalist moral code. From a position of moral neutrality, he moves to one of moral concern'.
69. Ranger (1972), p. 15f has pointed out the apparent flaw in Horton's Intellectualist Theory, viz. his failure to see that 'cults of affliction – which represent the *multiplication* of the idea of spirit rather than its increasing insignificance in face of the idea of God – were in themselves a response to nineteenth century "enlargement of scale" '. Horton has acknowledged the validity of this criticism in his revision of the theory: cf. Horton (1975), p. 221f, n. 6.
70. I would agree with Monica Wilson (1971), p. 11 who holds that 'the simple-minded interpretation of the Christian and Muslim missionary as simply a tool of imperialism, or economic expansion, is not tenable'.

CHAPTER 7

THE SHONA 'CHURCHES OF THE SPIRIT': PROMINENCE OF THE HOLY SPIRIT IN DOCTRINE AND RITUAL

In the preceding chapter, I have attempted to show that members of 'Churches of the Spirit' would seem to regard the Holy Spirit in the light of the Shona traditional territorial spirits, believing him to use their leaders and prophets as his mediums. Expatriate missionaries unwittingly facilitated this conceptual transposition of traditional notions upon the Holy Spirit, through their identification of the Biblical God Yahweh with the Shona 'high god', Mwari. As Gelfand points out, Shona converts to the Mission Churches soon found out that the Christian Mwari was silent, whereas they had been used to hearing their spirits speak.[1] This discovery had led to the earlier mentioned accusation against the Mission Churches, that they repressed the message of the Spirit. On the other hand, the fact that today among Shona adherents to the traditional religion, Mwari is commonly referred to as *Mweya Mutsvene*,[2] Holy Spirit, would seem to be largely due to the prominence which is given to the Holy Spirit in the doctrine and ritual of the Independent Churches.[3]
Speaking of Johane Maranke's African Apostolic Church, Aquina makes the doubtful statement that 'Christ forms the centre of Catholic and Protestant Christianity. He is also the model and guide of the Apostles'. She goes on:

> However, he has left the earth and returned to his Father in heaven, from where he sent men his Holy Spirit to remain with them. For the Apostles the Holy Spirit is, therefore, the most important person of the Trinity. He explains to them the Bible and all Christ taught . . . The Apostles, accordingly, centre their religion round the Holy Spirit . . . The Holy Spirit is believed to animate every Apostle and to give him power to preach the Gospel and to purify the country.[4]

Her charitable interpretation is, however, far too theological to satisfactorily explain the dominant position which the Holy Spirit holds in the 'Churches of the Spirit'. Considerably deeper insights are to be gained from investigating how certain basic Shona notions about their spirits have been transposed upon the biblical Holy Spirit in these Churches. This will be the subject of the present and subsequent chapter. Johane Maranke's calling, as it is narrated in the 'New Revelation of the Apostles', at first sight seems to be wholly conceived on biblical grounds:

> I was surrounded by a white cloud and the Holy Spirit descended on my head . . . I was singing the new song, "Hallelujah! Hallelujah!", and when I walked up an anthill a book was given me. Then my father and brothers came. They knelt before me and started worshipping God. . . . That is how the Holy Spirit first came to me.[5]

It contains, among others, elements of the narrative about Christ's baptism and transfiguration, as well as a reference to the dreams of Joseph in Genesis.[6] However, it will also be remembered that, traditionally, Shona people show signs of respect to all persons whom they believe to be possessed by a spirit. Similarly, during the annual meeting of all Apostles living in Johane's home area, he would lead his followers up Mt. Nyengwe and prophesy for them on personal and Church matters, much like a traditional spirit medium would do during the annual rain-making ceremonies.[7] As will be remembered, it was at Mt. Nyengwe that Johane had been first 'filled' with the Holy Spirit.

It has also been seen that contemporary Shona continue to carry out customary ritual, simply because their ancestors have told them to do so, although they often profess ignorance about the deeper significance of such rituals.[8] In like manner, Johane claims to have received the full charter of his Church by the direct inspiration and command of the Holy Spirit. According to the 'New Revelation of the Apostles', he was given two books, written in a foreign language, which contained all the laws and customs of the African Apostolic Church and which he was able to read through the mediation of the Holy Spirit.[9]

Members of the two major branches of the 'Churches of the Spirit' are even convinced that there are two separately identifiable kinds of Holy Spirit, the 'Spirit of Zion' and the 'Spirit of the Apostles'. According to Daneel, 'only the more sophisticated leaders recognize one and the same Christian Spirit, but distinguish its diverse activities within the context of different Church groups'.[10]

However, it can be reasonably assumed that the number of these 'sophisticated leaders' in the two camps is very small indeed, because soon after Johane Maranke had founded his own Church an agreement between Mutendi's Zionists and the Apostles already proved to be impossible. Shortly after starting his own movement, Johane Maranke considered joining Mutendi's Zionists. A delegation of Apostolic Church leaders was, therefore, sent to Mutendi. However, 'during the discussions the Spirits of the Zionists and of the Apostles came out and they were seen to be different'.[11]

Spirit Possession

In the early stages of their existence, all movements belonging to the 'Churches of the Spirit' were opposed to the western-style mission schools and made conscious efforts to lure children away from the latter. Thus Mutendi would adopt the following device:

> He sends one of his elders to a . . . place to let the people know he will be there on a particular night . . . He goes to this place and beats a drum. This is a sign to all the young people in the neighbourhood to proceed to where he is. On their arrival he . . . works himself up into a hysterical paroxysm of religious mania, throws himself about and makes noises like a baboon barking. After this he leads all the people up a hill. By this time they are all imitating Mutendi . . . These gatherings . . . can last for three nights.[12]

At the same time as Zionist leaders were telling people that 'education is not a good thing and nobody ought to go to school', and that those who had joined the Church of the White missionaries had 'the mark of the beast on their foreheads'; they themselves were opening numerous 'schools' of their own, although the administration gave official recognition only to schools which were under the supervision of Whites.[13] As is obvious from the above report, these Zionist 'schools' consisted of large gatherings of children and young people, during which there was 'much singing both day and night', while 'some of the young boys and girls became hysterical, throwing themselves on the ground in convulsions and making weird noises'. At the time, local Native Commissioners noted the similarity between this behaviour of participants in Zionist 'schools' and 'the hysteria practised in *kubatwa neshavi* ("beating or dancing the *shavi*"), but now referred to as *kubatwa noMweya* ("beating or dancing the Spirit")'.[14]

I have mentioned in the preceding chapter that the night after Johane Maranke had been filled with the Holy Spirit, he was seized by continuous spells of spirit possession. His elder brother Anrod relates that when Johane had come home in the evening he had a strange look on his face and kept saying 'Amen! Amen!' His family thought he had gone mad,

> but our uncle, Peter Mupako, who was a Zionist at the time, knew better, and assured us that it was not madness but the Holy Spirit that had taken hold of Johane. Mupako also fell down and was filled by the Spirit.[15]

Significantly, Gwati, like Simon Mushati one of Johane's maternal cousins, comments on the events in these words: 'Peter was a Zionist, and when he saw Johane he was filled by his [Johane's] spirit'.[16] In this connection, one should recall my contention that among the Shona, 'possession' is a well known stereotype and that during seances involving stranger-spirits it occasionally happens that even uninitiated spectators may spontaneously fall into a trance.

Mutendi, whose career as a Church founder started with spells of spirit possession similar to those which befell Johane Maranke, admitted that these at first 'confused' him, 'because the nature of such seizures were similar to the traditional *shavi* possessions'.[17] However, he soon became convinced that what he experienced was the Spirit of God acting in him. His first major sermon was addressed to a congregation of people who belonged to the Dutch Reformed Mission Church. As a result, 'many believers were possessed by the Holy Spirit'. The audience became frightened and some people ran away, saying

> That man arouses *zvitebwe*[18] . . . Some people laughed when they saw the others getting possessed by the Spirit . . . From that day, when he had preached and the people had laughed at the Spirit, he never stopped preaching the Word of God, and was greatly strengthened by that same Spirit.[19]

The reaction of some of the audience to the effect of Mutendi's preaching is still typical of the official attitude in the Mission Churches today. Thus Phillip Chikomboya, who is now a junior Apos-

tolic baptizer, was once 'seized by the Holy Spirit' during a Methodist service. As everybody in the congregation told him that he had been invaded by a demon, he left the Methodist Church and joined the Apostles.[20] However Murphree describes how the Budjga of northern Shonaland, who belong to the Methodist Church, have developed a popular ritual of *shavi*-exorcism which is regularly administered to numerous persons during their massive annual gatherings, as soon as their clergy have retired for the night.[21]
Aquina points out that the 'charism of tongues . . . is always the first charismatic gift which a new member of the Church of the Apostles develops'. Noting that this 'gift' forms part of the Shona's traditional religious stock-in-trade, she states laconically that 'among the Apostles the charisms of languages conferred by the Holy Spirit is usually the gift of the English language'.[22] Later in this chapter, further instances of traditional forms of 'possession' will be encountered in connection with my discussion of the various 'gate tests' which are customary in the 'Churches of the Spirit'. The subject of 'possession' will also recur in the following chapter, in connection with several forms of Zionist and Apostolic diagnosis and therapy.

Baptism in 'Jordan'

In Mutendi's own view, which is crystallized in chapter fifteen of his Church's Holy Book, *Rungano*, he spread a 'new message, as had never before been heard in Rhodesia'.[23] It consisted of two central elements, namely the need for each of his converts to be baptized in 'Jordan', as a pre-condition to beatification, and his subsequent possession by the Holy Spirit. As regards the second element of Mutendi's new message, its newness is doubtful when it is viewed against the evidence which I have just presented. However, even the 'Jordan' theme can be shown to fit into traditional Shona religious lore and practice. One need only remember the importance which the concept of the mythological pool (*dziva*) holds in the traditional religious thinking of Shona people. Power-conferring beings are believed to have emerged in the past from the *dziva;* mythological ancestors and powerful diviner-healers are thought to have acquired their ultra-human powers through entering into it; the *dziva* is also associated with the most powerful territorial spirits; submersion in a pool still forms part of the chiefly installation ritual in some Shona chiefdoms; and I have shown that the *dziva* theme has definite associations with the ritual reintegration of the ancestral spirits into the community of the living.[24]

The fact that Bishop Shoko, founder of the 'Zion Apostolic Faith Mission', should have had to wait eight full years[25] before the Spirit 'came out properly' in him, after his baptism in 'Jordan', is also a familiar notion among Shona traditionalists. A similar fate has befallen some well-known diviner-healers after they have completed their professional apprenticeship, and it will be remembered that not every person who has been indicated as a potential host to a stranger-spirit immediately succeeds in entering into his new role. The order in which the first followers of Johane Maranke entered into 'Jordan' established their seniority in the movement, is a fact worth recalling. Every pool or stretch of river which the Apostles or Zionists use as their 'Jordan' becomes for them the temporary abode of the Holy Spirit, and its waters are believed to cure sicknesses and to drive away demons.[26] This belief is underscored through the kind of songs which are sung during the baptismal ritual, such as: 'You have come today; cast away the spirit of Satan, which follows you'.[27] However, the Holy Spirit demands that each candidate, prior to his entry into 'Jordan', must make an exhaustive confession of his sins. As is the case during the Zionist 'clearing of the house' ceremonies on top of mountains, which I discussed in the preceding chapter, prophets guard the entrance to 'Jordan' and subject each neophyte to one or several diagnostic prophecies before allowing him to be baptized, or rejecting him as an unsuitable candidate.[28]

Prophetic activities: Christian substitute for diviner-healer practices?

During a survey conducted by Daneel, it was found that, apart from their healing activities, the biggest attraction of 'Churches of the Spirit' derives from their prophetic activities, such as the performance of miracles, the accurate prediction of a person's future, the detection of a married partner's infidelity, or assistance received in one's search for a suitable marriage partner or a better job.[29] This evidence is matched by the high incidence of the belief among members of these Churches that God sends us messages through prophets who are inspired by the Holy Spirit.[30] In this connection, it should be remembered that, of old, the Shona have been conditioned to hearing 'voices' and oracles through the medium of people whom they regard as being possessed by different spirits. For this reason, it is not surprising that contemporary Shona belonging to 'Churches of the Spirit' subject themselves so willingly to the searching questions of prophets.

Prophecies also occupy an important place in Zionist and Apostolic ritual. Aquina reports that during the weekly meetings of Apostolic congregations the sermons are always followed by prophecies which prophets and prophetesses address to individual members. It is interesting to note that whereas prophets address themselves equally to men and women, prophetesses carry messages almost exclusively for men, mostly in order to give them advice concerning their wives and daughters. Prophetesses tend to use their prophetic prerogative much as female members of stranger-spirit cults use their privileged status during their possession to bring pressure upon their husbands and male-dominated surroundings. Thus male interlocutors sometimes bring a prophetess's lengthy admonitions to an abrupt end by telling her for example:

> Stop now prophesying; we are not like diviners who never come to an end. We must close so that we have still time to heal the sick and to exorcise some *mashavi*.[31]

Marie-Louise Martin has claimed that in contemporary African prophetic and messianic movements 'the prophets and messiahs "possess" the "Spirit" like an impersonal power'.[32] In a similar vein, G. C. Oosthuizen has maintained that in these movements 'the Spirit has . . . become the monopoly of the leader [and] . . . is actually at his disposal'.[33] He sees the cause for this state of affairs in the fact that

> the functions of the ancestor spirits have been transferred to the Holy Spirit, or simply "The Spirit", so that in the independent post-Christian movements their "holy spirit" is no longer the Holy Spirit of whom we learn in Scripture.[34]

On the last five pages of his second volume of *Old and New in Southern Shona Independent Churches*, Daneel holds that the claims made by Martin and Oosthuizen 'are misleading and cannot be categorically applied' to the Shona Zionists and Apostles.[35] He emphasizes that 'a content-analysis of the diagnostic and therapeutical "prophecies" . . . [in the "Churches of the Spirit"] reveals continual references to God (Mwari) as the one who directs the observations of the prophets' and who is not conceived of 'as a force to be manipulated'.[36] However, he also says that

> one cannot deny the continual and sometimes one-sided concern of Zionist and Apostolic prophets with the spirit-world. Yet this preoccupation derives from their urge to deal effectively with the reality of a host of ancestral, alien and vengeful spirits, *as traditionally conceived of* . . .[37]

It is my view that in this last statement Daneel essentially concurs with the views expressed by Oosthuizen, above. It contains the admission that prophets in the 'Churches of the Spirit' deal with those powers which Shona people have traditionally conceptualized under the guise of different spirits, in the same manner in which their traditional specialists have dealt with them in the past. Whereas each of the traditional specialists had derived his ultra-human powers from his own spirit-familiar, prophets act by the strength of the one 'Holy Spirit', which, nevertheless, has been seen to be at least distinguishable as a separate 'Spirit of Zion' and 'Spirit of the Apostles'. The continuity in prophetic practice between the present and the past is also obvious from other statements by Daneel, such as his writing that in their struggle with the 'demons', prophets identify themselves with the Holy Spirit, speaking like the spirit-possessed medium of a territorial spirit; that people are aware of the fact that seemingly authentic possessions by the Holy Spirit may be cleverly staged by the prophets;[38] and that

> the ideal in the Shona prophetic movements . . . is . . . "to be seized by the Spirit" in order to reveal God's will . . . But . . . the temptation for the Shona prophet, Zionist and Apostle alike, is to utilize the authority through the Spirit for selfish purposes that are not aligned to God's Kingdom, to try to win or impress people through theatrical effect, sometimes at the expense of order during Church services.[39]

Through their prophets, the 'Churches of the Spirit' maintain a system of constant control over their members. Its particular aim is to keep witchcraft activities among them at a minimum. Apart from the initial check of all newcomers before their baptism in 'Jordan', members are regularly subjected to 'gate-tests' before they are allowed to participate in their Church's weekly or annual services.[40] During the first few weeks after their baptism, self-confessed or suspected witches have to endure unusually long and penetrating prophecies, which are intended to make them yield all their 'medicines'. While people readily admit minor transgressions of the moral code, they are naturally reluctant to confess the more serious ones. Known or suspected witches, adulterers, and thieves who try to pass the 'gate' undetected, find their way blocked by prophets whom the Holy Spirit causes to break out in violent bodily shakings, snorts, and loud shouts. Thus 'inspired' and 'acting like pagan diviners', they ask leading questions such as:

> "Did you not talk to this or that woman?" "Did you not get something from her?" "Did you not conceal it in your hut?"[41]

Daneel has devoted a whole chapter to describing the activities of some prophets who have gained renown as specialists in the prophetic treatment of one specific kind of disability, 'in much the same way as a *nganga* may build up a reputation for the successful treatment of a certain ailment'.[42] One of the followers of the *Shinga Postora*,[43] kraalhead Makuwise, says he joined the movement because its main protagonist, Elison Mutingwende, 'helped me when I was ill'. Elison removed from his house a horn and a crab-like object with human hair attached to it. He revealed to Makuwise that this harmful 'medicine' had been 'sent' to him by his relatives in Salisbury who owed him money. His eye-trouble was diagnosed by the prophet as being caused by the adulterous activities of his wife. He was also said to be under attack from his maternal grandmother's ancestral spirit, who wanted her 'grandmother's goat' which was owed to her by Makuwise's father. 'Long ago', Makuwise says,

> before Christianity came we were enlightened by our *nganga* concering family matters. So that is the most important thing in this Church . . . Everything is revealed to me through prophecy in the same way as [through] traditional divination.[44]

As a diviner-healer 'sees', with the help of his spirit familiar, the spiritual causes of his clients' troubles, prophets in the 'Churches of the Spirit' 'see through the Spirit' what other people cannot see, namely the hidden sins of converts and established Church members, as well as the 'medicines' and familiars of witches and sorcerers (*varoyi*).[45]

The 'Churches of the Spirit' as alternative communities

Daneel claims that the prophetic diagnosis in the Shona 'Churches of the Spirit' is a legitimate Christian substitute for diviner-healer practices.[46] However, in the same context he admits that 'it is doubtful whether all the prophetic diagnoses made in the name of the Holy Spirit can rightfully be ascribed to true revelation from this divine source', because 'prophets sometimes misuse the diagnostic spell by suggesting all sorts of mystical threats to their patients' in order to recruit them as new members.[47] In the light of this admission, Daneel's assertion that the prophetic diagnoses are a Christian substitute for diviner-healer practices stands on weak ground. He is certainly right in pointing out that in the 'Churches of the Spirit',

> instead of meeting the demands of the afflicting ancestral or other spirits through propitiatory rites, a deliberate stand is made against ancestor worship and traditional invocation of the spirits. Exorcistic treatment in the name of the Christian God or other symbolic activities, with the emphasis on the healing and protective power of God, is prescribed.[48]

However, in my opinion the use in these Churches of exorcism in the place of appeasement is a logical consequence of their posing as a new all-embracing family, into which new members enter through their baptism in 'Jordan'. They are made to believe that in this wider community they have found a complete substitute for those protective forces which, in their traditional past, had enabled them to enjoy an uneasy *modus vivendi* with the spiritual powers that be. They have transferred their allegiance from the traditional to new sources of power, but they have not changed their views about the intrinsic interaction between these powers. Their readiness to entrust themselves to new and seemingly more powerful 'guardians' and diagnosticians is not without precedents in Shona history. It has been seen that in the past Shona chiefs were eagerly attempting to avail themselves of what appeared to them as superior power in the first missionaries. I have also shown that in this society there exists a long-standing tradition according to which far-reaching changes in its circumstances are believed to be caused through the intrusion upon it of 'stranger-spirits'.

That the rejection of their traditional 'guardians' would be so radical today, as is evidenced by the Zionist prophets' referring to them as 'Satan' and 'demons', is understandable in the light of the profound changes to which Shona society has been subjected in its more recent past. Nevertheless, in contrast to Daneel, I argue that among Shona belonging to the 'Churches of the Spirit' the basic notions about power and powers, as I have outlined them in the first five chapters of this book, have not changed. To highlight my contention, one typical prayer of Elison Mutingwende exemplifies many similar ones:

> Hallelujah! in the name of the Father, the Son, and the Holy Spirit... Bu, bu, bu...[49] You are the God of this place of Mishek, Abednego and Daniel. You are the God of heaven. You, Jesus, we trust in your power... Bu-bu-Hosanna... We believe in you, Lord of Truth. Gather your people, Lord, the men and the women! Even the chiefs... have come awaiting you, Jehovah. They are awaiting you, Mwari, Lord of the [tribal] leaders. You are the God of the lame and the weak-bodied people – bless them with your power, Jesus. Amen... Haija! Hai-i-ya![50]

I have argued that the 'Churches of the Spirit' present themselves to their members as an alternative community, whose spiritual resources are more powerful than those which are at the disposal of the traditional Shona religionist. I further submit that the manner in which these new spiritual resources are thought to exist and act, in other words, their conceptualization, has not changed substantially against the background of traditional pattern. The 'Christianized' version of the 'bringing-home' (*bona*) ritual for deceased Church members is another case in point. According to Daneel, it 'has retained an essential function of the traditional ceremony', namely 'the belief that the living relatives of a deceased person have the responsibility to assist the spirit in acquiring its full ancestral status'.[51] He also says that the recruitment value of the adapted ritual should not be overlooked, and he notes the remarkable flexibility of Independent Church leaders when it comes to permitting officials and members of their Churches to participate in traditional rituals, such as the 'bringing-home' ceremony, which they condemn unreservedly in their official doctrine.[52]

It is clear that Daneel does not agree with those Zionist office-bearers who deny the existence of any resemblance between the traditional 'bringing-home' and the 'consolation ceremony' which they conduct for their deceased Church members. Nevertheless, he believes that the latter has a 'distinctly Christian character'.[53] The two main reasons for his opinion are, firstly, that the Zionist 'consolation ceremony' is directing the soul of the deceased to heaven and not back to its former homestead as is the case in the traditional 'bringing-home' ritual; secondly, that 'the ritual unit' responsible for accommodating the spirit of the deceased is no longer the deceased's kin-group, but 'a congregation which considers itself to represent the Christian Church'.[54] In contrast to Daneel, I would again emphasize that these two differences are a logical consequence of the desire, on the part of the 'Churches of the Spirit', to prize their living and deceased members out of their natural families and to incorporate them into the new, all-embracing 'family' as which they present themselves to their followers.[55]

The Spirit and singing

It has been seen that Johane Maranke claims to have received the full charter of his African Apostolic Church through direct inspiration by the Holy Spirit. This includes also several of the songs which are frequently used in this Church.[56] But apart from these, all singing which takes place during Apostolic ritual is thought to be 'a way of witnessing the arrival of the Holy Spirit'.[57] Daneel has observed the healing activities of Apostolic prophet Mabondo and his wife Febi at their healing centre close to the Apostolic Church headquarters in Maranke. He writes that as the patients pass the healers, Apostolic songs are continually sung. The same is true of mass baptisms organized by Apostolic prophets at the instigation of kraalheads, who in this way hope to rid their communities of witches. Amid the singing of Apostolic songs, every villager has to pass the 'gate' formed by the prophets. At the end of the ordeal, those who have failed to pass the test remain seated on the 'wrong' side of 'Jordan'.[58] Particularly gifted singers are distinguished by wearing the letters *Hari-*

kros[59] on their white tunics and it is their task to direct and sustain the singing during Apostolic services.

Apostles eschew the use of musical instruments, which they consider to be given over to the world and the powers of evil. Nevertheless, it is obvious that their rhythmic singing is similar in its effect upon the participants to the rhythmic percussion discussed in chapter four, in connection with the trance-state which is observable in the seances involving stranger-spirits. Jules-Rosette writes that

> the sign of God's presence among the singers is said to be the Holy Spirit as manifested in the visible gestures of waving the head and crying *Amen ! Jesu !* or other confessions of Christ. Sustaining the chant-like character of songs over a long period of time is also an evidence to members that the Spirit is with them. It is the feature of songs, particularly as sung by the *maha-rikros* in healing sessions but also sung in *kerek*, that Murphree refers to as the almost 'hypnotic' quality of Apostolic singing.[60]

Another interesting point is made by Aquina. She notes that women members of the African Apostolic Church are under no circumstances allowed to preach during services. However, male members of the Church told her that 'when the women sing they preach'. What this means is that when the women grow tired of listening to a preacher, or if they disagree with the content of his sermon, they start singing and gradually raise their voices to such a volume that the preacher is forced to stop.[61] This is another obvious instance suiting Roger Gomm's definition of spirit possession as 'a set of strategies for redefining situations'.[62] Under the protection of their supposed possession by the Holy Spirit, they are able to exert pressure upon their menfolk, just as prophetesses have been found to do under similar circumstances.

NOTES

1. Gelfand (1973), p. 133.
2. Hannan's *Standard Shona Dictionary* (1974), gives the following meanings for *mweya:* 1. Air; 2. Breath; 3. Vapour; 4. Spirits; 5. Soul; 6. Disposition; 7. Smell. The word belongs to the third noun class which has been seen in chapter two to be associated a.o. with atmospheric phenomena, things growing up like a tree, etc. The adjective, *tsvene*, stands in Chikaranga and Chizezuru for the following English words: Right (as it should be); Sound (in

good condition); Acceptable. Hannan states that among the Karanga, Manyika, and Zezuru, *mutsvene* has been used to translate the English word 'holy'.

3. Cf. Daneel, (1971), p. 85.
4. Aquina (1967), p. 208.
5. Daneel (1971), p. 316.
6. Cf. Mt. 3, 16; 17, 5; Gn 37, 10.
7. Cf. Daneel (1971), p. 336.
8. Cf. above, chapter three.
9. Cf. Daneel (1971), pp. 317, 321.
10. *Ibid.*, p. 329.
11. Thus Simon Mushati, Johane's maternal cousin and life-long right-hand man. Cf. *ibid.*, p. 328.
12. Report of the Native Commissioner of Bikita district, August 1931. Cf. *ibid.*, p. 419.
13. *Ibid.*, p. 415. In 1929 an investigation revealed that there were seven schools whose teachers acknowledged Mutendi as their head, in the following districts: Bikita (4), Ndanga (2), Gutu (1). As Mutendi was not the only Zionist leader opening schools, the very existence of the schools run by the Dutch Reformed Church and the American Board Mission was threatened. In fact, in several cases attendance at the mission schools became so depleted that they had to be closed down: Peaden (1970), pp. 58, 62.
14. Daneel (1971), pp. 415, 418. One of the accusations which were brought against Mutendi and others who had opened illegal Zionist schools was that in these schools children were taught in the native tongue, i.e. in Chikaranga or in Chindau. Peaden (1970), p. 64 reports that in the Chipinga area Zionists urged parents to send children to their schools where they would be taught only in Chindau, and he adds that 'the word Chindau means not only the language but the whole culture and custom of the people'.
15. Daneel (1971), p. 319.
16. *Ibid., loc. cit.*
17. *Ibid.*, p. 290.
18. Vengeful spirits which cause destruction among the cattle of their enemies under the guise of maneless lions.
19. *Rungano*, p. 17f. Cf. Daneel (1971), p. 294.
20. Daneel (1974), p. 28.
21. Murphree (1969), p. 75f.
22. Aquina (1967), p. 209. Among the more intelligible examples of such speaking in tongues is the following: 'Brrr; viner, viner, Christ diviner, Christ diviner; wonderful Christ diviner. Wonderful gloria, tora [take] picture, half, half'.
23. Daneel (1974), p. 189.
24. Cf. above, pp. 31, 72, 78, 115.
25. Cf. Daneel (1974), p. 187. In Daneel (1971), p. 292 Shoko is quoted as having said: 'I had to wait a full year before the Spirit entered into me. I often dreamt that I was filled by the Spirit, but it would not "come out" (*buda*) properly'.

26. Aquina (1967), p. 213.
27. Daneel (1974), p. 275.
28. Cf. Daneel (1974), pp. 275, 293.
29. *Ibid.*, p. 181.
30. *Ibid.*, pp. 142, 185.
31. Aquina (1967), p. 213.
32. Martin (1964), p. 161, quoted in Daneel (1974), p. 347.
33. Oosthuizen (1968), p. 124, quoted in Daneel, *op. cit.*, p. 348.
34. *Ibid.*, *op. cit*,. p. 129, quoted in Daneel, *op. cit.*, p. 350.
35. Daneel (1974), p. 348.
36. *Ibid.*, p. 349.
37. *Ibid.*, p. 351. My emphasis.
38. *Ibid.*, p. 351f.
39. *Ibid.*, p. 352f.
40. Cf. *ibid.*, pp. 292f, 297, 300.
41. Aquina (1967), p. 211. Cf. also Daneel (1974), pp. 300, 303.
42. Daneel (1974), pp. 261-92.
43. Shona name meaning 'Courageous Apostles'. This movement specializes in removing malignant 'medicines' from people and their homesteads.
44. Daneel (1974), p. 276.
45. Cf. *ibid.*, pp. 271, 297. One of the Zionist leaders warns that when the *varoyi* give them their familiars, hyenas and owls, to burn, these can act like gunpowder and cause a loud explosion.
46. Cf. *ibid.*, pp. 336-8.
47. Daneel (1974), p. 337.
48. *Ibid.*, p. 338.
49. Elison speaking in tongues.
50. Daneel (1974), p. 269.
51. *Ibid.*, p. 320. As for the traditional form of the *bona* ritual, cf. above, chapter three.
52. Cf. *Ibid.*, pp. 117, 121-39. In his description of a Zionist 'consolation' (*runyaradzo*) ceremony, Daneel notes that soon after the burial of the deceased had taken place, the customary delegations were sent out to determine the cause of death. They consulted both traditional diviners and Zionist prophets: *ibid.*, p. 128.
53. *Ibid.*, pp. 121, 324.
54. *Ibid.*, pp. 125, 320-5.
55. It would lead too far to discuss in detail another point made by Daneel (*ibid.* p. 326f) in this connection, but it deserves to be mentioned. In his view (cf. Daneel [1971], p. 94f) – which shows some similarity to the presentation in Mbiti (1969) – ancestors act as mediators between the living and the high god Mwari. Therefore he notes with surprise that no mention was made during the Zionist 'consolation ceremonies' of the mediating role of the deceased. Instead, 'the focus was on the mediating role of the Church'. This is precisely on what the focus lies during the traditional 'bringing-home' ritual also, i.e. on the mediating role of the officiants, and not on that of the ancestral spirit who is to be accommodated. This is a further proof that the basic pattern of

the traditional 'bringing-home' ritual has been preserved in the 'Christianized consolation ceremony'.

56. Daneel (1971), p. 320.
57. Jules-Rosette (1975), p. 154.
58. Daneel (1974), pp. 211, 279, 303. On page 289 Daneel very aptly refers to these mass baptisms as 'baptismal ordeals'.
59. This is Jules-Rosette's spelling. She thinks the term is derived from 'High-Cross': cf. *op. cit.*, p. 153. Aquina (1967), Murphree (1965 & 1969), and Daneel (1974) each use different spellings of the term.
60. Jules-Rosette (1975), p. 164. Her reference is to Murphree (1969), p. 107. *kerek* = religious service.
61. Aquina (1967), p. 206. Cf. also Jules-Rosette (1975), p. 155.
62. Cf. above, chapter four.

CHAPTER 8

THE SHONA 'CHURCHES OF THE SPIRIT': SPIRIT-CONTROLLED DIAGNOSIS AND THERAPY

Dreams have been seen to play an important role in the relationship between traditional Shona and the various spirit beings they believe in. Diviners, witches, and spirit mediums are liable to receive their calling by having persistent dreams about the spirit who has chosen them as his host. Diviner-healers are often thought to be warned in their dreams about people who are on their way to consult them, and in this way to gain a fore-knowledge of their particular problems. Many of them claim to learn their healing arts entirely from their indwelling stranger-spirits, who show them in their dreams herbal concoctions and the places where they are to collect the ingredients for these.[1] Not surprisingly, therefore, Daneel found Shona members of both Mission and Independent Churches to be equally convinced that God sends messages to men in their dreams.[2] I have already related the decisive vision of Johane Maranke, through which he recognized his calling to found the African Apostolic Church.[3] However, long before this he had had similar visions. From the age of six, Johane regularly saw 'many people on a cloud in the east'. Missionaries of the American Methodist Mission who heard of his dreams and visions took them as a sign that Johane would one day become a minister in their Church.[4]

While such sympathetic interpretation of dreams to this day remains a rare exception among expatriate missionaries, Shona prophets in the 'Churches of the Spirit' will invariably approach dreams in the traditional manner. Thus, before he founded his Zionist Apostolic Faith Mission, Bishop Andreas Shoko dreamt that he was climbing a high mountain. As he reached the summit, he suddenly came across a troop of baboons. When these saw his white face, they were so frightened that they jumped over a ledge and fell to their death. The prophets told Andreas his dream signified that he would become filled with the Holy Spirit – 'hence the white face!' – and then drive the evil spirits out of many people.[5] In like manner, Mutendi received his calling to found the Rhodesian branch of the Zion Christian Church through a series of dreams, and his doubts about which of the two South African Zionist Churches he should join were also resolved through a dream and its subsequent prophetic interpretation.[6]

Daneel recognizes that in the 'Churches of the Spirit' the revelations of the Holy Spirit 'reflect some affinity with traditional divination conditioned by *shavi* spirits' and that for this reason the diagnosis of the Spirit-inspired prophet is very similar to that of the traditional diviner-healer. However, he claims that 'the Apostles and Zionists distinguish clearly between the Holy Spirit (*Mweya Mutsvene*), the angels (*angerosi*), and the ancestral spirits (*midzimu*)', because

> the Holy Spirit speaking through the prophet ascribes illness to the action of spirits, but he *never* reproaches people for neglecting the ancestors! Neither do the *angerosi*, when they visit people in dreams.[7]

For Zionists and Apostles, Daneel says, the Holy Spirit and the *angerosi*, whose visits they receive in their dreams, are 'good spirits'. These spirits call them into Church membership and encourage them to remain steadfast in their new faith. The *angerosi* replaces the ancestor and urges the dreamer to fulfil his obligations towards what I would call his new 'kin-group', i.e. the Church.[8] However, members of 'Churches of the Spirit' are still frequently visited in their dreams by their ancestral spirits, although they classify these as 'bad spirits', 'whose requests must be evaded, if possible'.[9] As the Apostle Sauro Garanuwako says:

> Once I dreamt that my mother's spirit came and gave me thick porridge (*sadza*) to eat. I was greatly disturbed and asked the prophets what this meant. They told me not to worry because it was just an ordinary visit with no implications. I thought it might be the beginning of ancestral trouble in my family in order to make us worship the ancestors . . . My mother was not inducted into the spirit world after her death, but should she start troubling us she might succeed in overcoming us and thereby cause us to accommodate (*kugadzira*) . . . her spirit.[10]

Prophetic diagnosis

In the preceding chapter, I dealt extensively with the prophetic activities of Zionist and Apostolic Church officials. The reason for returning to this subject once more is to demonstrate its close connection with the healing activities which constitute the major attraction of the 'Churches of the Spirit', as I mentioned in the same chapter.[11] In Daneel's view faith-healing as it is practised in the 'Churches of the Spirit' highlights more than any other practice 'the confluence of old and new'. This is so because,

> based on a rejection of divination as practised by the *nganga*, it nevertheless concerns itself with the very forces which the *nganga* regards as the cause of ailment. The Holy Spirit, too, indicates that ancestral spirits, witchcraft or other personal forces are the main source of affliction. In other words, *the Prophet recognizes and deals with the psychological causes of the patient's trouble in terms which are familiar to the latter. Therapeutical treatment however, reveals to what extent the prophet dissociates himself from the type of solution prescribed by the traditional doctor.* Instead of advising ancestral propitiation to regain the protection of the guardian spirits, liberation from the afflicting agent through the healing power of God is emphasized. Magical medicaments and charms are supplanted by objects symbolizing the curative powers of the Christian God.[12]

Two important claims are made in this statement which are, however, not borne out by Daneel's copious material: firstly, that prophets reject the divination practices of diviner-healers and, secondly, that their therapeutical treatment of patients is essentially different from that practised by diviner-healers. Against this view, I reiterate my earlier view that these 'differences' derive from the fact that the prophets regard themselves as 'spirit mediums' and 'diviner-healers' possessed by the Holy Spirit, whose aim it is to cajole their clients into membership of the 'Churches of the Spirit'.

A characteristic example of prophetic diagnosis is contained in the case history of a Shona Methodist called Miche, who suffered from persistent spells of mental disorder.[13] These were diagnosed by several diviner-healers as being caused by Miche's paternal grandfather who was said to have turned into an avenging spirit (*ngozi*). When Miche found himself unable to meet all the requirements prescribed by the diviner-healer to appease the avenging spirit, he consulted several 'Cord-wearing Zionists' and Apostolic prophets. He says that these prophets told him 'the same things as the *nganga*'. Besides, they all urged him to join their Churches. He was told that if he would join, they would sprinkle his homestead with water and 'fence it off' from danger with their pegs.[14] These measures were to assist him to pray with 'power'.

Miche refused to meet these conditions, but, driven to desperation, he soon afterwards travelled to Mutendi's Zion City for treatment. There he was placed under the care of the paralyzed prophet Nison Mutuwira. Assisted by his interpreter, the evangelist Simeon, Nison put himself into a prophetic trance with the following prayer: 'God of Enginasi and Moyo,[15] I thank you this morning. Help me, as I am a person of the flesh, I ask power from you. Let your spirit come on me!' Then he spoke in tongues: 'If we justa to see a text ret to wit. I see . . . Hosanna! Moriah, Moriah[16] . . . for we just to see mistake . . . rindi ta . . .' Simeon gave Miche the explication of the prophecy:

> God says you are being troubled by an evil spirit of your own family. This spirit is very strong and said something about cattle. There is also another spirit which works together with this one. The big one is from your dead father. You sometimes dream of him speaking to you.[17]

Miche agreed that this was true. Then there followed a series of further prophecies:

> God says you once went to a *nganga* who lives near this place. At present the sons of your sick relative are talking to a [magical] basket because they want all the evil spirits to turn their attacks on you. Because of this you always have dreams and become very weak . . . "Go and tell this story to my servant [i.e. Mutendi] and I will drive the spirit away through his hands", says Mwari. When you have told my servant the whole story he will give you water and papers that have been prayed for. When you have the water you must go to your home and call the evangelist living nearby. He will sprinkle water in your house . . . After this, you and your family will live in peace, but remember, every member of your family should be sprinkled with water. Your wife must drink some of the blessed water for her stomach troubles, and so must the others. If you do not give heed to what I say, your whole future will be ruined, says the Lord. You will be without a job and walk around like a mad person.[18]

The similarity between Nison's diagnostic method and that of a diviner-healer divining with the help of his stranger-spirit, without the aid of dice, is obvious. After a short prayer, the Holy Spirit manifests himself[19] through the bodily tremors of the prophet, his snorts, and his speaking in tongues. Like a diviner-healer, the prophet is supposed to be unaware of what the Holy Spirit reveals through him amid bursts of 'speaking in tongues'. Therefore, like his traditional counterpart, he needs an interpreter who explains the content of the diagnostic spells of prophecy to the clients. Nison knew Miche's case to the last detail. This is far less surprising than it would seem at first sight. Before his case came up for diagnosis, Miche had already spent two days in Zion City. Besides, he had already consulted so many diviner-healers and prophets, that by the time he came to see Nison his troubles were well known beyond his immediate home area. However, even where a prophet does not have such an obvious advantage as this, by probing in the manner of a diviner-healer for the background factors in the life of his client, he is soon able to become specific in his diagnostic analysis.[20] It has been seen in the preceding two chapters that members of 'Churches of the Spirit' are constantly exhorted by their prophets to make exhaustive confessions. Likewise, during the diagnosis of a client's problem, prophets need not regard it below their professional honour to make a statement like that of prophetess Asalina Shuvai, who admonished a woman with the words: 'God says you must tell me the truth about this case, because I am only a woman myself and I do not want to lie'.[21]

Daneel's categorical statement in the above quotation that prophets, 'instead of advising ancestral propitiation to regain the protection of the guardian spirits', emphasize 'liberation from the afflicting agent through the healing power of God', is also hardly borne out by the facts, as I have already mentioned. Whenever prophets come to the conclusion that illness or misfortune have befallen a client because of the legitimate claims which a deceased ancestor or his living representatives have against him, they use the traditional standardized patterns of conflict as an explanation and recommend the appeasement of the irate ancestral spirit as the only solution, exactly as any traditional diviner would do.[22] It is not surprising, therefore, to read Crawford's claim that some diviner-healers have even adopted the divining methods used by the prophets of Independent Churches.[23]

The 'Man of God's' healing power

In chapter six I have described how Mutendi gained far-reaching influence in Rhodesia through his alleged ability to grant fertility to barren women and to send rain to parched lands. However, the single most important factor in the rapid growth of his popularity was his apparently *nganga*-like power to heal the sick and to exorcise evil spirits. At the beginning of his career as a Church founder, Mutendi was constantly out on campaigns, calling on his fellow Africans to be converted. The *Rungano*, the ZCC's Holy Book, claims that his preaching was accompanied by astounding miracles, through which many people's health was restored and a few were even raised from the dead, for

> it has not been written [in the Bible] that now that Jesus preached with great force many people became converted because of his preaching, or that they followed him because they wanted to hear the teaching of Mwari, but that many people followed him because they wanted to be healed.[24]

Later the throng of sick people and barren women who sought his help increased to such a degree that Mutendi had to convert his Zion City into a virtual healing centre. Increasingly, he would depend on a number of reliable resident prophets, like Nison Mutuwira, to pay attention to the less seriously ill, while he himself treated the more critical cases.[25] The confidence people have in Mutendi's healing powers is vividly reflected in the sermon which was held by an evangelist at the centre:

> This place is like Bethsaida. Just as the power, in olden days, came through great personages such as Abraham and Jacob, it now derives from the God of Mutendi and Enginasi Lekhanyane. *Mutendi's power is stored away like the grain of a farmers' cooperation.* He can cure you of all ailments. Assuredly he is the Man of God![26]

In the sermons of Mutendi and other ZCC officials, Mwari is constantly represented as the God of Power, whose gift to 'live in strength' is mediated through Mutendi to his followers. According to Daneel, due to an ever-increasing demand, Mutendi's power to heal became his most important manifestation of power.[27] There are scores of testimonies from people who claim to have been healed by Mutendi, whom his followers respectfully refer to as the 'Man of God'. To many of them the 'Man of God' appeared in their dreams, bathed in radiant light, and sometimes touching them with his holy staff. Most of them joined the ZCC after their cure, some because they feared their illness would return unless they did so. Others say they themselves have been endowed with great power to heal ever since they have become members of Mutendi's Church.[28] Both Mutendi and Johane Maranke have cleverly used such enthusiastic witnesses to their extraordinary healing powers as propagators of their Churches. Thus it is said that it was part of Johane's expansion strategy to ordain as leaders people who had been healed and to send them back to their homes to win converts for the new movement.[29] During their prayer sessions, members of the Zionist Women's Association have referred to Mutendi as their 'very own *nganga*'.[30] One may be tempted to say that this comparison of Mutendi with a diviner-healer was merely inspired by the praying women's exuberant sense of gratitude to their healer and that it should not be taken too seriously. However, the fact is that there exist many parallels between the healing practices of Mutendi and other leaders or prophets in the 'Churches of the Spirit', and those of traditional diviner-healers. Most members of these Churches state that they 'were made to join through illness', and cases are on record in which prophets declared that they had to baptize an entire family in order to obtain the power to pray for one of its members.[31]

Just as in the past one would go from one diviner-healer to another in search of a cure, many people today try out various 'Churches of the Spirit' in search of prophetic healing or fertility. They retain their membership in a particular Church only for as long as they are satisfied that it is 'a true Church', because it maintains their health. Quite obviously, such people regard baptism in the light of a traditional 'medicine'. Leaders in these Churches complain that individuals who are in need of a job, a cure, protection against the spirits and wizardry, etc., often come to be baptized, but remain as members only until the specific objective is achieved.[32] Remarking upon the Apostles' large turnover in membership, Aquina notes that 'those who are not cured miraculously . . . leave the Church in disappointment'. Local congregations consist of a core of three or four men who have been Apostles for the last ten or twenty years of their lives, whereas the majority of Church members form a 'floating fringe' staying for a couple of months and then reverting to the religion from which they first broke away'.[33]

It is hard for one to avoid the impression that Daneel, who is investigating the possibility of opening a theological training centre for leaders of the Independent Churches in Rhodesia,[34] deliberately refrains from making statements which could be taken amiss by these leaders and jeopardize the implementation of his plans. For this reason, his argument oscillates between statements which are only just short of the accusation made by members of Mission Churches, namely that Mutendi was 'nothing but a great *nganga*', and the attempt to deny any connection between the prophetic healing through the Holy Spitit and the healing techniques of diviner-healers. However, the bulk of his material proves, by his own admission, that there are 'direct parallels' between the prophetic treatment of patients and the medicinal practices of diviner-healers.[35] Besides, the fact that as early as in 1646 Shona rulers who had become subject to the Portuguese, and whose territory had seen a brief spell of missionary activity, needed 'baptismal water' for their accession rituals, shows the remarkable ability of traditional institutions to absorb newly arriving phenomena without themselves changing in essence.[36] The 'baptismal water' was obtained from the Portuguese stronghold at Sena and was drunk by the new ruler only after it had been 'medicated' by a senior medium of a territorial spirit.[37]

Symbols of power as 'medicines'

For Zionists, it has been seen, power was stored away in Mutendi like grain is in a farmer's co-operative. Patients seeking a cure for their ailments at the founder's Zion City or through the mediation of a ZCC prophet elsewhere, are obliged to make use of the 'medicines' which symbolize Mutendi's healing power and have passed through his hands for his blessing. The most frequently used objects are:

> Mutendi's staff as the most potent expeller of spirits; sanctified water which must be drunk, or which, used to spray people, doorposts and living rooms, drives out spirits or protects against wizardry; sanctified needles, worn by healers on the lapels and

> used at the appropriate time as 'injection' instruments, e.g. by the constant pricking of any part of the body or nasal cavity for the removal of 'evil' blood or water; sanctified newspaper, torn to shreds and burnt under specific circumstances, e.g. for inhaling of smoke in cases of pectoral complaints or to drive out *shavi* spirits threatening to possess a child; sanctified strips of linen cloth, worn by women around their abdomen to stimulate fertility. The prophetically prescribed number of knots in it is supposed to determine the sex of the unborn child. Worn by children immediately after birth or by grown-ups around their necks, arms, and legs, these strips serve to protect the body against visiting spirits; sanctified soap is used for cleansing the body.[38]

Discussing the official ZCC view, according to which these objects are 'primarily the visual symbolic concretization of the Divine Power', Daneel recalls that the beer pot which symbolizes the presence of the deceased ancestor during the traditional 'bringing-home' ritual,[39] is at times addressed as if it were the ancestor in person:

> The pot becomes the ancestor! This is also true of Zionist healing practices. The healing hand of the 'Man of God', the sanctified staff and other objects prescribed by the prophets are not only representative symbols, as might be inferred from discussions; at a given point they become the very power of God.[40]

Sacred staves play an important role in all 'Churches of the Spirit'. Like Mutendi and Johane Maranke, also every 'Cord-wearing Zionist' archbishop has a staff which his followers believe possesses higher curative power than those of his subordinates. The staff used by a subordinate prophetic healer only becomes really powerful after the main leader of the Church has laid his hands on it or has touched it with his own staff.[41] As regards the holy water which is extensively used in the course of each prophetic therapy, its power is compared by an Apostolic prophet in his sermon with the instant availability of the electric current:

> If we take water and pray for it, we just sprinkle it in our houses as a poison to the *varoyi* and it blocks their entry. Last night I met two *varoyi* at my house. When I sprinkled holy water on them they fell down on the floor, powerless. The power of God is like the electricity used by *changamire* [reference to author]. You must put your hand at the correct place and the power immediately becomes operative.[42]

Daneel draws attention to the fact that some Apostolic healers continually emphasize the need for patients to have subjective faith in their treatment. However, it would be wrong to interpret this emphasis as an outflow of their Christian convictions. The two well-known Apostolic healers Febi and Mabondo, whom I have mentioned in the preceding chapter, often scold their patients because these fail to confess all their sins and refuse to bring all their 'medicines' and 'things of the world' to have them burnt. 'As a result', the healers complain, 'all our praying and our sprinkling with holy water will be rendered useless'.[43] Behind these words stands the same philosophy of power and powers which has been found to be characteristic of the traditional diviner-healer in chapter five. The same is true of the large-scale use by Zionists and Apostles of vomiting as a means to remove from persons evil shadows or substances with which they are claimed to have been bewitched.[44]

The hold which the traditional conceptualization of powers still has over members of 'Churches of the Spirit' is perhaps best exemplified by the dilemma in which a Zionist who is believed to have been bewitched by a fellow Zionist finds himself. When Mutendi was still alive, Zionist witches were said to use water which had been blessed by Mutendi to get at other Zionists, sprinkling it as they went to the house of their victim. The 'spirit of Mutendi' was then believed to open the door, allowing the witch to enter the house and to carry out his nefarious deeds. From this it is obvious that Mutendi's power, stored in him like grain in a farmers' co-operative, could be used for good or for evil. He had no ultimate control over its use. This casts the same shadow of ambiguity over him and his power as has been seen to be characteristic of the traditional Shona diviner-healer.

Healing centres

Even the idea of a healing centre is not unknown to the traditional Shona past. According to Daneel, all well-known diviner-healers have one or more spare huts at their homesteads in which their patients can stay if they should require prolonged treatment.[45] In this way a small healing centre has also grown up at Johane Maranke's homestead. Instead of encouraging the growth of one large 'hospital' like Mutendi's Zion City, Johane has established a cluster of healing centres in his home district, each of which is under the supervision of a healer. 'Much like the *nganga*', many of these healers become known for their successful treatment of specific ailments. Thus individual patients arriving at Johane's place can be directed to healers who are most likely to be able to help them.[46] Apostolic healers live with their patients in an intimate face-to-face relationship and much of their time is spent in ritual activity. There are lengthy prayer meetings in the morning and late afternoon, during which the healer exhorts his patients to confess all their sins, lest the holy water and the prayers lose their power and the treatment be rendered useless.[47] Although the ZCC also maintains a network of smaller healing centres in outlying districts, all serious cases are sent to the 'hospital' at Zion City for treatment. On his arrival there, a patient is welcomed by the 'doctor in charge' who first of all sprinkles him with holy water. Then his complaints are noted down and he is given a slip of paper with the name of the prophet who will diagnose his illness at a later stage. Mutendi has laid down that all high-ranking ZCC office-holders from outlying districts have to pay regular visits to Zion City where they spend at least two weeks ministering to the needs of the patients.

Every non-Zionist patient arriving at Zion City is regarded by the community as a potential convert. When Mutendi was still alive usually the prophets would suggest to the patient that he should stay for a few days in order to come into direct contact with the 'Man of God'. The prophets also leave no doubt that the best chance for a patient to be cured lies in his undergoing purification in 'Jordan'.

Each day, after morning prayers, the 'Man of God' used to bless the supplies of water which patients had brought for their treatment during the day. Patients who had special requests were then summoned to an interview with Mutendi. It ended with Mutendi's laying his hands on them or with a brief touch by his holy staff. All other patients would receive every morning the laying-on of hands from the officiating Zionist ministers. After this they would spend long hours with the prophets in whose care they had been placed These would then go with them through the processes of diagnostic prophecies and 'medicinal' treatment which I described earlier in this chapter.[48]

Fear of witchcraft and the 'Churches of the Spirit'

So far I have mentioned only in passing that belief in witchcraft does play a role in the 'Churches of the Spirit'. Thus it has been seen that one of the main objectives of the pre-baptismal prophecies is to bring to light any secret attachment of the neophytes to wizardry. It will also be remembered that after Johane Maranke's death his maternal cousin Simon Mushati was accused of having caused the Church founder's death through bewitchment. However, witchcraft occupies such an important place in the Spirit-controlled diagnosis and therapy in these Churches that, in the remainder of this chapter, I will have to deal almost exclusively with this subject.

The intensity of the belief in witchcraft among the prophets themselves can be gauged from the kind of statements they make in public. Thus, for instance, before opening the proceedings for a 'baptismal ordeal', which lasted from 10 a.m. until sundown, the prophet Mufeyi alleged that two women had visited the hut in which he was sleeping during the night to bewitch him. Their aim was to prevent him from demasking them on the following day. However, through the Holy Spirit he had been able to fight them off.[49] Another preacher, on a similar occasion, harangued the suspected witches in an attempt to prepare them for entering into 'Jordan':

> Have your meat supplies run out, that you want to kill your children and eat their flesh? Do you want us to subject you to a poison ordeal . . . ? . . . It is only a fool who goes about naked. A good person obeys the law [of this Church]. You women from Mukungu, you take your witch medicines with you to sow disunity in the Church. When the sun went down last night we saw an owl which glowed like fire and we killed it. So, in what did you profit from it? Have we not destroyed many of your devices?[50]

Nearly all members of Zionist or Apostolic Churches accused of witchcraft are females, and far more women than men are stopped at the 'gates' leading to the grounds in which the annual *Paseka* ceremonies take place.[51] This is well in keeping with the traditional use in Shona society of witchcraft accusations as a strategy of attack, and of witchcraft confessions primarily as a means of enhancing one's status. A particularly enlightening case is that of the younger daughter of Zionist Bishop Kuchekenya. After her puberty she regularly received call-dreams from her maternal grandmother, who had been a notorious witch. She was, therefore, subjected to exorcisms during each Sunday service until the dreams ceased to trouble her. Eventually she became a prophetess of standing in her father's congregation.[52]

Speaking of the 'Courageous Apostles' (*Shinga Postora*), an Apostle of Johane Maranke's African Apostolic Church has made the remark: 'Perhaps this new Church has no witches. In ours there are many!'[53] This admission is accentuated by the claim of the above mentioned Apostolic healer, Febi, that she was blinded by the medicine which an Apostolic woman in a neighbouring village had used against her because she was jealous of Febi's success and prestige.[54] Similarly, holy water blessed by Mutendi, being one of his Church's most widely used conveyor of the 'Man of God's' healing power, has been seen to turn into dangerous matter in the hands of Zionist witches. Worse still, Mutendi himself is known to have had three women in his own household who were considered to be witches. They were his older wives who were past child-bearing age and who were known to be envious of his younger wives. These women were not allowed to enter Mutendi's living quarters, which amounts to the 'Man of God's' acknowledgement of the fact that even he had to live in fear of possible bewitchment. Two of the three had confessed attempts to bewitch their husband's favourite wives, while the third had lost all her children and was held responsible for the death of one of Mutendi's wives. To save face, a private 'gate-test' was conducted in Zion City for the leader's women and children by two reliable acquaintances in the vestry of the church, out of sight of all other Church members.[55]

Officially, the reason for the fact that known witches are not chased away by their Zionist or Apostolic husbands and that they do not lose their membership in the Church is that chasing them away would be 'contrary to the new law of love'.[56] However, the true reason for this lenient treatment seems more likely to be that husbands whose wives have been diagnosed as witches regard their regular subjection to prophetic scrutiny as an effective means to keep their alleged nocturnal activities at a bearable minimum.[57] This interpretation is supported by the fact that often women who are suspected of practising witchcraft are forced by their husbands to join a 'Church of the Spirit'. Aquina, therefore, suggests that husbands may even employ the services of prophets to settle tensions in their families. For the purpose, such conflicts are often expressed in terms of dreams. A husband may give expression to his fears by narrating dreams in which he claims to have seen his wife leaving the house during the night. A prophet can then take up the lead and start diagnosing the case through the aid of the Holy Spirit, forcing the woman into a confession.[58] Daneel has found that the Zionist version of the traditional rain rituals, described in chapter six, is used for the same purpose even by non-Zionists. At the behest of their husbands, or of their own free will, suspected witches queue up to pass the 'gate' to the mountain, and the prophet's reaction either confirms their professed innocence or vindicates their husband's fears and suspicions.[59]

That the members of the 'Churches of the Spirit' view witchcraft basically along the traditional lines, is further emphasized by the frequency with which Zionist and Apostolic prophets diagnose bewitchment as the cause of illness. The sick register of the 'hospital' at Zion City classifies the complaints of persons arriving for treatment as light ailments, barrenness, inability to find a suitable partner for marriage, stomach-ache, 'evil spirits', headache, and swelling of the limbs.[60] Of these, the last four are almost exclusively attributed to bewitchment. Daneel found that nearly half of the 373 cases entered in the register for 1964 were diagnosed as having been caused by wizardry.[61] Treatment of these conditions ranges from the drinking of holy water blessed by Mutendi to the burning of 'sanctified newspapers'.[62]

As has been seen, in all 'Churches of the Spirit' it is accepted that persons who are believed to engage in witchcraft activities cannot always be cured and must, therefore, be tolerated. This is not surprising, if indeed witchcraft allegations pertain to a 'set of definable social niches', as shown in chapter five, and therefore reflect endemic areas of conflict which often cannot be radically changed. Nevertheless, in their official doctrine these Churches uphold the principle that all 'evil spirits' – and this includes, of course, witchcraft-inducing stranger-spirits – can be exorcised through the power of the Holy Spirit. Among the Apostles, writes Aquina, 'the charism of exorcising evil spirits is, therefore, regarded as the greatest of all the charismata because it can purify the country from evil . . .'[63] Johane Maranke, in his call-vision, was promised that he would be able to exorcise any kind of stranger-spirit, and during the first baptism

ceremony of his Church he is reported to have driven many evil spirits from people as they were passing through 'Jordan'.[64]
During or at the end of Zionist and Apostolic services, prophets regularly subject women who confess to having had call-dreams from witchcraft-inducing stranger-spirits to treatment by exorcism in an attempt to prevent them from becoming confirmed witches. According to Daneel, 'special songs' are sung to bring on the troubled person a spell of spirit possession, 'whereupon the prophets drive the spirit off with the laying-on of hands and the sprinkling of blessed water'.[65] During an exorcism observed by Aquina, an evangelist brought on the 'spell of spirit possession' by rubbing the head of the girl from whom a stranger-spirit was to be exorcised, first gently, and then ever faster. A prophet danced over her, and eventually the bishop commanded the stranger-spirit to leave the girl.[66]

Witch-hunting prophets

I noted in chapter six that some secessionists from main-stream 'Churches of the Spirit' have turned back to traditional forms of worship. Others have become well-known witch-hunters whose methods are thoroughly traditional, but may also include some features which have been inspired by their contact with Christian forms of worship.[67] Constantly on the move and always faced with the risk of being prosecuted for their activities, they are regularly summoned to distant villages to conduct large-scale witch-hunts, in the name and through the power of, the Holy Spirit.
After his arrival in the village, the witch-hunting prophet spends the next few days collecting a small fee from all adult villagers, interviewing them for hours to learn their current views and suspicions. Then follows the 'baptismal ordeal' which usually lasts for the best part of a day. Dressed in a long white robe, the witch-hunter stands knee-deep in the water at the point where the people have to enter 'Jordan'. Each person passing the prophet makes a public confession of his sins, while the village elders and kraalheads listen intently. Often the prophet brings on himself spells of prophecies to elicit further hidden sins from those whom he suspects of being guilty of theft, adultery, and especially witchcraft, stopping them if they refuse to respond to his promptings. Those who have passed the prophetic 'gate' are baptized by assistants amid the singing of

Apostolic songs, and led to the opposite bank of the pool or river. In the end, those who have to remain seated on the 'wrong' side of 'Jordan' are declared to be the witches.[68]

Crawford has pointed out that the Shona have not developed large-scale witchcraft-eradication movements as they have sprung up in neighbouring countries to the north of Zimbabwe, in answer to the suppression of the traditional poison ordeal through the former colonial administrations. Among the Shona, the role of these movements is largely fulfilled by the 'Churches of the Spirit' and by the roving, witch-hunting *nganga*-prophets.[69] Witchcraft-eradication movements and their Shona counterparts both suffer from the same dilemma: they cannot keep their promise of bringing about 'instant millennium'[70] once and forever through cleansing communities of witchcraft and its evil concomitants. Without elimination of the evils of disease and death, or of social conflict and frustration – all of which for the Shona are evidence of the practice of witchcraft – no 'eradication of witchcraft' can endure.[71]

Exorcism of avenging and ancestral spirits

My argument in chapter three, that an avenging spirit (*ngozi*) is the conceptual objectification of a person's or a family group's state of being in debt, is vindicated by the blunt admission in the 'Churches of the Spirit' that it is virtually impossible to rid a person of an avenging spirit without his first paying the traditional *mutumbu*, i.e. a young unmarried woman or twelve head of cattle, to the wronged party.[72] Guilt cannot be 'driven away', unless justice is done first. Persons who are unwilling to appease the avenging spirit through the payment of *mutumbu* may, however, seek to escape the spirit's attacks by meticulously adhering to the laws of a 'Church of the Spirit'. This can be interpreted as a conscious effort on their part to rid themselves of any responsibility towards their family group's liabilities and obligations, by demonstratively entering into allegiance with the Church as a substitute 'family', whose powerful protection they hope will safeguard them against further attacks from the avenging spirit. Through this bit of legalistic acrobatics, the Church member is thought to have escaped the rightful claims of the avenging spirit – however, only for as long as he obsequiously fulfills the injunctions of his new 'family':

> In our family . . . a *ngozi* spirit . . . has killed many of our kinsmen. My father's younger brother has already paid ten oxen to the children of the evil spirit in order to appease it . . . If I had not joined this Church I should also have had to pay *mutumbu*. The *ngozi* is constantly on the prowl . . . At the least sign of a blunder in my church life . . . it could unleash its wrath on me. The [*ngozi*] . . . is . . . an old woman whom our grandfather chased away . . . and who died alone in the bush. *Only in this Church do I find protection against the ngozi.*[73]

While it is thus admitted that Church members can hardly evade the necessity of settling their debts with an avenging spirit in the traditional manner – fleeing to the 'safety' of Zion City being a rather extravagant alternative, which only the few will be able to make use of – the 'Churches of the Spirit' officially feel confident about their ability to contain the power of 'normal' ancestral spirits. They use this term only with regard to the spirits of those deceased who have failed to receive a place in heaven and who have, therefore, turned into 'evil spirits' (*mweya wakaipa*). From their abode in the 'pit of fire', they are said to be occasionally allowed to visit their living relatives, 'in an attempt to find rest'. They are also believed to have the power, as had the angered ancestral spirits of old, to 'open the door' of a house for witches to enter and do harm to its inhabitants. Zionists and Apostles as members of their protective Church 'family', defend themselves against these evil spirits by taking countermeasures, such as sprinkling blessed water or burning strips of 'sanctified' paper, which are believed to drive them away. However, things are complicated by the fact that even the spirits of believers who have gained a place in heaven, and who are thought to act as guardians vis-à-vis their descendants by interceding for them with God, are believed to be able to bring about disaster in their families by asking God to punish the disobedient.[74]

NOTES

1. Cf. above, chapter five; Bourdillon (1976), pp. 175, 190f, 205. Daneel (1971) p. 148.
2. Cf. Daneel (1974), pp. 142, 185. The question was: 'In what ways does God send us messages?' The answer 'Through dreams', was chosen by 50% of the Dutch Reformed, 46% of the Roman Catholic, 54% of the ZCC members,

50% of the *Ndaza* Zionists, and 48% of the Apostles. The percentage becomes even higher (viz. 78%) for members of 'Churches of the Spirit', if one adds the number of those who chose the answer: 'Through prophets inspired by the Holy Spirit'. This is justifiable, according to Daneel, because 'a clear distinction is seldom made between a vision (*kuona* = to 'see') and a dream (*kurota*) and [because] the Spirit is believed to manifest himself to prophets through both'. Cf. *ibid.*, p. 142.

3. Cf. the beginning of the preceding chapter.
4. Daneel (1971), p. 316.
5. Cf. *ibid.*, p. 292f. The traditional meaning of the white colour, as symbolizing the presence of spiritual beings, their power, as well as health and fertility (cf. chapter three, *passim*) is amply documented among the 'Churches of the Spirit' also: cf. e.g., Daneel (1971), pp. 290, 298, 316, 322; (1974), pp. 143ff, 158, 235.
6. Cf. Daneel (1971), pp. 289f, 298. Daneel (1974), p. 143ff records also numerous cases of people who claim to have had call-dreams which left no doubt about which Church they were to join and whose dreams, therefore, needed no prophetic interpretation.
7. Cf. Daneel (1974), p. 333f. His emphasis.
8. Concerning this transfer of loyalties from the family group to the new and wider Church 'family', cf. my argument in the preceding chapter.
9. Daneel (1974), p. 334.
10. *Ibid.*, p. 334. *Kugadzira* is the term used by the Southern Shona for the 'bringing-home' ritual. Zion Christian Church minister Samson Bracho states that his ancestors visit him in his dreams and that their visits forebode illness. If, however, the Church founder Mutendi appears in such a dream, standing in the middle, illness is averted. 'If he does not stand in the middle, I will certainly fall ill. If I dream of my *midzimu* without seeing Mutendi, I have been warned that physical affliction is coming my way'. Cf. *ibid.*, p. 159.
11. Fifty to sixty per cent of Mutendi's and Maranke's followers were prompted to join their Churches because of the diagnostic and therapeutic work allegedly performed by the Holy Spirit through the prophets. Cf. Daneel (1974), pp. 181, 186.
12. Daneel (1971), p. 463. My emphasis. Cf. also Daneel (1974), p. 217ff.
13. Cf. Daneel (1974), pp. 245-53.
14. Details about this type of protective treatment which is regularly administered by diviner-healers have been given in chapter five, above.
15. Enginasi Lekhanyane was the founder of the Zion Christian Church in the Transvaal: cf. chapter seven. Moyo is Mutendi's clan name: cf. Daneel (1971), p. 287.
16. ZCC members refer to Mutendi's Zion City also with the Biblical place name 'Moriah'.
17. Daneel (1974), p. 249. Concerning the prophet Nison, cf. also *ibid.*, p. 193.
18. *Ibid.*, p. 250.
19. Cf. *ibid.*, p. 214. Significantly, Daneel writes that the Holy Spirit manifests 'itself'. Note the similarity to the ease with which experienced diviner-healers are able to enter into the trance state.

20. Cf. Daneel (1974), pp. 215, 223f.
21. *Ibid.*, p. 222.
22. *Ibid.*, p. 226f.
23. Crawford (1967), p. 204. As regards divination of witchcraft through prophets belonging to what he terms 'Pentecostal Churches', cf. *ibid.*, pp. 205, 210, 221-43.
24. *Rungano*, 46, 6. Cf. Daneel (1971), p. 297.
25. Cf. Daneel (1974), p. 189.
26. *Ibid.*, p. 206f. Daneel's emphasis.
27. *Ibid.*, p. 255.
28. Cf. Daneel (1974), pp. 159, 193 and *passim*.
29. Daneel (1971), p. 330. Cf. also Daneel (1974), p. 190.
30. Cf. Daneel (1974), p. 258: 'First I had no child, but now there are children. When Maria walked to the grave, she sang: "Jesu wangu [My Jesus] . . .". In like manner, we must honour Moyo, for many women found help in that Church. Many of them had no children. Nor did they have goats with which to pay the *nganga*. Today, however, we have our very own *nganga*, that is Moyo. Let us all run to Moyo, for with him there is deliverance'. The woman refers to Mutendi under his clan name, Moyo.
31. *Ibid.*, p. 195f. 'I became a Christian through the illnes of my stomach', runs the testimony of a junior evangelist in Johane Maranke's Church of the Apostles: *ibid.*, p. 196.
32. Daneel (1974), pp. 17, 20, 68, 195. Cf. also the end of subsection *Secessions*, in chapter six.
33. Aquina (1967), p. 207f.
34. Cf. legend on front flap of dust jacket of Daneel (1971).
35. Cf. Daneel (1974), p. 232f.
36. Cf. also Schoffeleer's evidence in chapter four, above, p. 92.
37. Cf. Bourdillon (1976), p. 25. From the beginning, Portuguese missionary activity in Mozambique had a rather sporadic character and was often linked with the interests of the Crown. Schebesta (1966), p. 65 writes that baptism was regarded by the African 'converts' as a rite which made them into 'Friends of the Portuguese' and entitled them to maintain free trade relations with the foreigners. A typical example is the baptism in 1560 of 'king' Gamba of Tongue (alias Otongwe), whose residence was some distance from the coast near Inhambane. The head of a Kalanga clan, he had been driven from his territory in Shonaland and established himself among the Thonga by force of arms. Thus he had additional reason to look for friends. Fr. Goncalo da Silveira S. J., who was soon to be killed in the course of a similar mission at the court of the Monomotapa, baptized Gamba after a mere three weeks of irregular instructions – through an interpreter – about the Christian doctrine and way of life. Soon afterwards four hundred of the 'kings's' subjects were also baptized. However, as soon as the festivities were over and the missionaries' presents exhausted, 'heathendom returned with all its brutality'. *Ibid.*, p. 64. Cf. also Theal (1907), p. 306-13.
38. Daneel (1974), p. 234.
39. Cf. above, chapter three.

40. Daneel (1974), p. 233.
41. *Ibid.*, p. 236.
42. *Ibid.*, p. 271. *Varoyi* = witches or sorcerers.
43. *Ibid.*, p. 235. Note also Febi's recipe for ending a woman's barrenness (*ibid.*, *loc. cit.*): 'While the *nganga* in the neighbourhood "clothe" the barren woman with medicine and the European doctors clean their wombs to prepare them for conception, I only need my red [linen] rope and the water that has been prayed for. I give this water to barren women to drink, after which it keeps their wombs in the correct position to receive the male sperm. I use the red rope to cast out the interfering evil spirits. Having tied up the woman I wash her again and again with holy water in order to wash away all the demons that try to keep her captive. Then I ask God to grant the woman a child. If she confesses all her sins and really believes, she will have a child'.
44. Cf. *ibid.*, pp. 220, 235, 252, 281. On p. 263 Daneel records the vision which the Apostolic witch-finder Jaka Mukurumbera had at the time of his baptism. He saw a pig which was vomiting white medicine and he was told that this was the 'big satan which vomits sickness onto people'.
45. Daneel (1971), p. 145.
46. Cf. Daneel (1974), pp. 190f. 327.
47. *Ibid.*, p. 210f.
48. Cf. Daneel (1974), pp. 201-9, 214.
49. Cf. *ibid.*, p. 283.
50. *Ibid.*, p. 295. Concerning the owl as witch-familiar, cf. above, pp. 111, 160, n. 45. Cf. also Daneel (1974), p. 271, where an Apostle interprets the trials which Christians are bound to endure in terms of witchcraft attacks. The Christian's response to these attacks is to be the same as it had been in the traditional past, viz. migration to another place of residence.
51. Cf. *ibid.*, p. 261, and Aquina (1969), p. 128. It should be mentioned that, due to Mutendi's renown as mediator of fertility, his Zion Christian Church has a particularly large female membership: *ibid.*, p. 67. From the beginning, Zionist preaching and healing activities proved to be especially attractive to women, leading to the accusation that Zionists abducted women by force. Cf. Daneel (1971), pp. 294, 435.
52. Cf. Daneel (1974), p. 298. The author promises to present the case in greater detail in a subsequent volume. For the moment, his significant hint that the girl was the bishop's *younger* daughter must suffice.
53. Daneel (1974), p. 272.
54. *Ibid.*, p. 211.
55. *Ibid.*, p. 301f.
56. *Ibid.*, *loc. cit.*
57. Cf. also *ibid.*, pp. 294, 304.
58. Cf Daneel (1974), pp. 161, 215f, 294; Aquina (1969), p. 128f. In an earlier article, Aquina notes that after every religious meeting of the Maranke Apostles which she attended, several women were exorcised, most of whom were young married mothers. She believes that the women voluntarily submitted to these exorcisms in order to be publicly certified by the Apostles to be free from witchcraft, 'so that no future accusation from fellow villagers and

affines could blacken their reputation in a village in which they lived as strangers'. Cf. Aquina (1968), p. 52.

59. Daneel (1974), p. 113.
60. Cf. *ibid.*, pp. 217f, 222f, 226.
61. Cf. *ibid.*, pp. 202, 205.
62. *Ibid.*, p. 217f. A typical instruction runs like this: 'You must pray with vigour, else you will die. You must drink the water which Mutendi has blessed with his prayers. He will press his staff against your stomach and give you a newspaper which you must use at home'. Cf. *ibid.*, p. 217. Other objects used to ward off witchcraft are holy cords and sanctified strips of linen. Cf. *ibid.*, pp. 218, 221f.
63. Aquina (1967), p. 214.
64. Daneel (1971), p. 324.
65. Daneel (1974), p. 298. Cf. also Murphree (1969), p. 75. Concerning the 'special songs', cf. subsection *The Spirit and singing*, in the preceding chapter; also Martin (1971), p. 115.
66. Aquina (1967), p. 215. Cf. also Daneel's (1974), p. 239-43 excellent treatment of these exorcist activities in the 'Churches of the Spirit'.
67. For detailed case histories, cf. Daneel (1971), p. 332; (1974), pp. 262ff, 280ff. Cf. also Gelfand (1971), with its revealing title. Hemans (1970), and Robinson (1971) portray what are essentially witch-hunting *nganga* who, however, declare themselves to be Christians. In connection with the last-mentioned author, cf. also Redmayne (1970).
68. Daneel (1974), p. 278f.
69. Cf. Crawford (1967), pp. 278, 283ff. In fact the famous *Mchape* witchcraft-eradication movement did reach the Shona, as Crawford recognizes. Cf. also Bourdillon (1976), p. 225f. Obviously what Crawford terms the 'Pentecostal Churches' are identical with the 'Churches of the Spirit'.
70. Cf. Willis (1970).
71. Cf. Bourdillon (1976), p. 227.
72. Cf. Daneel (1974), p. 254.
73. Statement made by ZCC minister Jacobo Matevure at Zion City. Cf. Daneel (1974), p. 244. His emphasis. Cf. also Miche's case which I have described earlier in this chapter.
74. Daneel (1974), p. 118. Cf. also Apostle Garanuwako's statement at the beginning of this chapter.

CHAPTER 9

SHONA SPIRITS: CONCEPTUALIZATIONS OF POWER

In my introduction to this book I said that, with Bourdillon, I view the Shona 'Spirits' as symbols of power, i.e. as conceptualizations, through which man endeavours to account for the manifold 'powers' he experiences as impinging on himself and the world around him.[1] I also made it clear that this does not mean that the individual Shona is conscious of the fact – as it appears to the anthropologist – that spirits are conceptualizations. What the anthropologist refers to as 'symbols of power' are for the Shona awe-inspiring, real spirits. Anthropological theory about symbolism has not yet solved the apparent contradiction between the assertion that these conceptualizations are thought to have been consciously constructed, and that they are, nevertheless, acquired by the individual member of a society by way of a largely unconscious process. For lack of a better expression, it is, therefore, still useful to refer to them, as does Durkheim, as 'collective representations'[2] and to make the matter-of-fact statement that these are handed down from one generation to the next.

Unified symbolic system

However, on what grounds can one regard what I have called 'conceptualizations' as being close to Durkheimian 'collective representations'? My argument has been clarified by Cohen's formulation that

> symbols are systematized together within the framework of dynamic ideologies, or world-views, in which the symbols of the political order are integrated with those dealing with the perennial problems of human existence . . . These two symbolic complexes support one another within a unified symbolic system.[3]

My having viewed the Shona 'ideology' as a unified symbolic system, in which symbols of the political order are integrated with those dealing with the perennial problems of human existence, has resulted in a wealth of evidence confirming Cohen's observation. Thus, for example, the territorial spirit medium's 'setting the heavens right'[4] involves his concern with disputes about succession to the chieftaincy as well as with procuring sufficient rain for the lands and protecting the crops against pests. Similarly, the ancestral spirits have been seen to uphold traditional norms with regard to property and authority, while the sanctions they are believed to employ against deviators, belong to the realm of 'the perennial problems of human existence', namely, death, illness, and misfortune. In Tables I-IV, I have systematically set out the spirit concepts resulting from the Shona's unified symbolic system.

In spite of its usefulness in so far as it has inspired the trend of my argument, Cohen's conception of a unified symbolic system suffers from lack of clarity on one important point. According to him, this system is made up of two symbolic complexes, one consisting of symbols of the political order, the other of symbols dealing with the problems of human existence. However, as I have shown, it is not merely *symbols* belonging to these two realms of human experience that are integrated into a unified system. Chiefs, spirit mediums diviner-healers, and persons accused of witchcraft activities – all of which can be said to belong to the 'political order'[5] – are real people, not just symbols. In the same way the perennial problems of human existence – death, illness, misfortune, and the illwill of one's neighbours – are experiential facts, not mere symbols.

Metaphysics

What then has been found to be unified in the Shona world-view is not merely a collection of symbols belonging to these two realms, but a uniform principle of power which is thought to be at work behind the phenomena belonging to both of them. Power is wielded both by tangible persons and by invisible entities, which latter are however, no less real an experience than the former. It is this experience of the all-pervasiveness of power which has led to its symbolization. As a symbol, its distinguishing characteristic is its endowment with personal traits, whether it is thought to be the property of certain people, or to manifest itself in the form of spirits, or to adhere to inanimate 'medicines'. Power figures in this unified symbolic system as metaphysics in the original sense of the word,[6] i.e. as a principle or entity which is operative beyond or behind the visible nature of things.

TABLE I

	Experience of Power in the perennial problems of human existence	Constant threat of drought; lack of fertility in lands, man and beast; pests; famine and disease			Powerful chiefs & spirit mediums are successful in procuring rain, stopping pests, ending infertility in men and beasts.
CONCEPTUALIZATION OF 'POWER'	**in chiefs**	Land in chiefdom is life & strength for its people. The chief, as present 'owner of the land', is the guardian of the chiefdom's life and strength.		Chiefs crave for the 'power' vested in the invading Whites' superior control over nature. Their subjects see the spirit mediums as more immediate representations of the powers which in their experience are least controllable.	
	in territorial spirits (*mhondoro*)	The original 'owners of the land' – i.e. the chiefdom's founder-ancestors and subsequent past chiefs – retain guardianship over the land and its people. Disasters are seen as punishments sent by them.	Territorial spirits threaten usurpers of chiefship with death	Important territorial spirits threaten White intruders with destruction, claim suzerainty over them.	The original 'owners of the land' possessed prodigious powers. Often these resulted from their having broken socially approved norms & their use of powerful 'medicines'.

in spirit mediums	The chief's investiture with the original founder-ancestor's power over the land and its people is dramatically expressed in installation rituals involving the mediums of the territorial spirits.	Mediums of territorial spirits uphold threat of their alleged control over rain and pests. As spokesmen of chiefdom's founder-ancestors, they claim the sole right to the appointment of new chiefs. 'Possession'!	Anti-White resistance is fomented in the name of territorial spirits' title to ownership of land. Whites have 'spoilt the land', having alienated large parts of it from its original owners, appointed chiefs of their own liking, and legislated against age-old customs of the land.	
Experience of power in the political (social) order	The chief, as 'owner of the land' wields power over his subjects, allots to them land, which they may occupy and cultivate.	The complicated rules of succession lead to frequent struggles between contending candidates for chiefship.	The chief's power is qualified by his more or less firm control over land and people in his chiefdom, also by his firm or weak character. Since the wide-spread occupation by Whites of Rhodesia, the chiefs' power has shrunk drastically, while that of the spirit mediums has suffered less severely.	Chiefs, and people aspiring to chiefship or spirit mediumship, are known to dabble with charms and 'medicines'. Those wielding power in society often dare to break social norms and seem to gain extraordinary strength thereby.

TABLE II

Experience of power in the perennial problems of human existence	Man's fate is unpredictable: Death, illness, and misfortune strike suddenly and indiscriminately. At his death, man has to enter into the unknown.			
CONCEPTUALIZATION OF 'POWER' **in ancestral spirits** (*midzimu*)	*Midzimu* have entered the realm of nature and acquired power over it. They thus belong to a field of experience both frightening and benevolent. They are in a state of 'hotness' until the successors to their rights to authority and property have been duly installed. Only the *midzimu* of people who enjoyed power vested in authority and property during their lifetime are feared.	*Midzimu* & *ngozi* always unleash their harmful powers against the wronging party's property and against his status of authority. However, man can accommodate himself to these erratic 'powers' by meeting their demands and thereby appeasing them. But in practice, people often play a 'wait-and-see game' with the powers.	In Shona lore the laws of nature, including death, drought and their accompanying misfortunes, are overcome by mythical heroes. Imitating important aspects of the heroes' actions in their 'bringing-home' ritual, Shona express their belief that man can prevail against the 'powers' which are at the root of the perennial problems of human existence. The expected result of performing the ritual is the family's return to normality through the deceased's 'resurrection' in the heir of his authority and property.	The 'coolness' of a composed temper is conceptually linked with the whiteness of drought-breaking rainwater, famine-ending flour, and with other articles used during the rituals of appeasement. The 'hotness' of an angry mood is conceptually linked with the red or black colour of articles figuring in appeasement rituals. The red/white symbolism appearing in Shona rituals of appeasement expresses their desire to return to the desired state of 'coolness' = normality. It also gives expression to their sense of ambiguity vis-à-vis their recently departed.
in avenging spirits (*ngozi*)	Some people die in an aggrieved state because they were wronged in their rights to authority and property. Maternal *midzimu* are believed to become par-			

Experience of power in the political (social) order	Acquisition and possession of authority and property are basic values to Shona people and determine a person's status in their society. At the death of a holder of authority and property, there is danger of quarrelling among his descendants over succession rights.	Holders of authority and property may treat their subjects and dependants harshly. However, in the long run, submitting to authority leads to one's own assumption of power. Sometimes one gets away with disregarding the others' status of authority or with illegally holding on to others' property.	The unity between members of the agnatic family and its alliance with the 'wife-giving' family is put to a hard test at the death of an important member. Claims to his authority and property may come from many contenders and can only be sorted out some time after the deceased's death.	The power vested in a person holding authority and property is ambivalent. The desirable state of its holder is 'coolness'.
	Women, living as 'stranger wives' in their husbands' lineage, are often deprived of their rightful due.			

TABLE III

Experience of 'power' in the perennial problems of human existence	Mild attacks of mental disturbance and hysteria; Frustration with one's non-influential social status; Puzzlement with, or envy over extraordinary giftedness and skill in others; Need of self-consolation over one's own lack of special skills.				
Conceptualization of 'power' in shavi spirits	Enlargement of scale		Strategy for re-defining situations		
	The slain invaders of Shonaland died in a country alien to them, where no 'bringing-home' ritual could be performed for their spirits. These would, therefore, seek to 'come out' in Shona people, who refer to them as *shavi*, 'stranger-spirits'. By extension, any foreigners, such as wandering traders or White colonials, who met their death without receiving a proper burial can turn into	The stranger spirit concept enables the Shona to accommodate newly arriving experiental facts in their traditional view of the world by interpreting these in terms of invading alien configurations of power. In *shavi* possession cults a dramatic stereotype of the different invading powers is created, and acted out through rituals of spirit possession.	The experience that some people excel in hunting is another aspect contributing to the stranger spirit concept. Successful hunters are believed to be possessed by the spirits of famous hunters who lived in the past. By extension, any person endowed with extraordinary skills is liable to be thought of as being host to a stranger spirit.	Under the protection of 'possession' by a stranger-spirit, persons can – express inclinations they would normally have to suppress; – find gratification from the fact that for a short period they are the focus of attention for a potentially large crowd of people; – openly manifest oblique aggressiveness (e.g. against a husband or men	'Possession' as a conceptual stereotype becomes familiar to children from an early age. 'Dissociation' as an expressional stereotype is susceptible to conscious manipulation by the 'possessed' subject, and through auditory and tactile stimuli. Every Shona is familiar with the notion that 'a spirit will always have its way': it is best for one to accommodate oneself to the powers that be. Call-dreams and physical disorders indicative of a

	[illegible] unmarried Shona for whom as a rule no 'bringing home' ritual is performed. The 'Baboon *shavi*' concept crystallizes the experience of man's ability to trespass across conceptual and physical frontiers.			– enjoy membership of a 'club', with frequent meetings and good food; – temporarily escape from the drudgery of everyday reality; – enhance their public image, and if they have attained senior status, earn moderate sums as fee for the initiation of new members.	are also stereo typed.
Experience of power in the political (social) order	In the course of their history, Shona have experienced several invasions by foreigners. Some of the invaders were killed in battle, others died of natural causes in Shonaland.	The vast changes in the political, social, and economic field which have resulted from the influx of White settlers into Shonaland have led to recurrent outbreaks of *shavi* possession cults, involving ever new kinds of 'stranger-spirits'.	People differ from each other through their possessing of or lacking special skills. Those who have special skills are more influential in society than those who haven't. In a traditionally male-dominated society, dynamic Shona women in particular are searching for means to enhance their status in public.		

TABLE IV

Experience of 'power' in the perennial problems of human existence	Death, illness, and misfortune strike suddenly and indiscriminately. People suffering from mild attacks of mental confusion or from persistent illness and misfortune; Strange, perverse, or overbearing behaviour of marginal and underprivileged people in society; Recurrent outbreaks of discord in a community.		
CONCEPTUALIZATION OF 'POWER' **in witches** (*varoyi*)	The witch is the paradigm of anti-social behaviour. His disregard for socially approved norms is believed to stem from his being possessed by a witchcraft-inducing stranger-spirit which endows its host with extraordinary powers to harm his fellow men. Witchcraft beliefs are a rationalization of suspicions and tensions in a community, which are, however, based on the experience that there *are* evil-minded people who shun no means to achieve their own ends, viz. to enhance their social status and to increase their wealth. Women are feared as	'Medicines' of both witch and *nganga* are conceptualized along the same lines as are the 'powers' visualized as spirits. The witch's 'medicines' are sent like 'invisible missiles', 'poisoning' the intended victim. Only way to gain reprieve from the attacker is through appeasing witch or launching a counter-attack with the help of a *nganga*.	Ordeal is the only means to prove beyond doubt one's innocence of witchcraft suspicions.

	[illegible] witches. Through veiled threats and public confessions to being possessed by witchcraft *shavi*, barren or unhappily married women can promote themselves to powerful positions in their communities. Affinity between spirit possession and witchcraft concepts: both can be used as strategy of attack and are sometimes combined.			
in diviner-healers (*nganga*)	'Black *nganga*' launch witchcraft attacks against people. Often they are 'graduates' from spirit possession cults.	People are able to acquire 'good' and (particularly from 'black *nganga*') 'bad' 'medicine' from diviner-healers. Guided by their *shavi* spirits, *nganga* must provide clients with 'medicines' to counter witches' 'medicines'. *Nganga* may also attempt to expel a witchcraft *shavi* from its host through vomiting and/or public confession.	The *nganga's* 'kingpin' position in Shona cosmology Divination: Only he is able to discover the ultimate cause of people's illnesses and death. His dice 'see' the truth and his *shavi* guides him in his diagnosing while he is in ritual trance. Before making diagnosis, the *nganga* probes into the client's private life, 'goes round the client's spirits',	Healing: Illnesses are 'sucked out' of clients' bodies. Treatment against infertility and impotence in man and beast, epilepsy, etc. Driving out of people avenging and stranger-spirits. 'Doctoring' homesteads against lightning and witches. *Nganga*'s *shavi* shows him (in dreams) remedies to be used for treatment.

		Protective 'pegging' of homes against witchcraft attacks.	and seeks to discover tensions between the client and his neighbours, particularly those stemming from infringement of others' rights to authority and property. The *nganga* tells the client which spirit wishes to 'come out' in him or what a particular ancestral spirit wants of him. Stereotyped call-dreams and physical disorders are indicative of 'calling'.	
		What is 'good' protective 'medicine' to the one, is 'bad' medicine' = witchcraft to the other who wishes to trespass against him. Both witches and *nganga* are believed to be able to 'raise' an avenging spirit and launch it on the attack.		
Experience of power in the political (social) order	In Shona society there exist endemic areas of conflict: feuds occur between rival contenders for the chiefship; quarrels break out between family members over rival claims to a deceased's rights to authority or property; women living as 'stranger wives' in their husbands' patrilineage are often accused of furthering the ends of their own lineage, thereby threatening the solidarity among the members of the former. Some self-assertive, ambitious personages dare to go against customary social norms, doing 'what others are		The existence of diviner-healers in Shona society since a long time ago is a power in the social field to be reckoned with. While people may be aware that *nganga* use in their divination whatever prior knowledge they may have of their case, and while an individual *nganga's* diagnosis may be rejected by the people, their faith in the divining process itself is unshaken. The same applies to their belief that only *nganga* can heal certain ailments which Western medicine seems unable to cure. Witchcraft accusations which are upheld by a *nganga's* diag-	

TABLE V

TRADITIONAL CONCEPTUALIZATION OF POWER	CONCEPTUALIZATION OF 'POWER' IN THE 'CHURCHES OF THE SPIRIT'
Powerful founder-ancestors, as territorial spirits, guard the chiefdom's life and strength, appoint chiefs as 'owners of the land', and 'come out' in spirit mediums. (Compare Table I)	The Church founder is 'owner of the job' and receives his extraordinary powers through possession by the Holy Spirit, who 'comes out' in him as 'Spirit of Zion' etc.; power is 'stored away in him', but he has no ultimate control over it and it is, therefore, amenable to unlawful use in witchcraft practices; (at the beginning of the founder's career) the Holy Spirit frequently seizes him and his followers in *shavi*-like manner; the 'gift of tongues' is highly regarded and a must for all prophets in the Churches; the founder's following is growing and held together by his prodigious rain-making, divining, and healing powers; he gains an aura of invincibility by denouncing the territorial and other traditional spirits as 'Satan', which action prompts the latter to accuse him of 'spoiling the land'; leaning on his extraordinary powers, he builds up healing centres and institutes rituals to counter the traditional spirit mediums' influence over chiefs and people. Secessions frequently occur over questions of rank and authority, when claims are often based on revelations by the Holy Spirit and witchcraft allegations are levelled against rivals.
Although conceptually there is no difference between the powers supporting the activities of diviner-healers and witches, Shona trust that *nganga* are able to diagnose the cause of their misfortunes and to provide them with the means to counteract attacks against themselves by the ultra-human agencies. However, the *nganga's* position in Shona society is highly ambiguous. (Compare Tables III and IV)	Posessed by the Holy Spirit, prophets regularly subject Church members to 'gate tests', involving exhaustive confessions, the principal aim of which is to detect and control witches; among the latter, women are in the majority; some prophets specialize in witch-hunting. Though people are aware that possession by the Holy Spirit may be faked, the prophets' services are used by husbands suspecting their wives to be witches and by women who wish to prove their innocence of such suspicions. Prophetesses use their prophetic prerogative as a strategy against male domination much as female members of traditional *shavi* cults do during their alleged possession. Women accused of witchcraft, after prophetic treatment, may themselves become prophetesses. Exorcism

seances are regularly held at the end of services, but it is admitted that not all witchcraft *shavi* can be driven from their hosts and witchcraft attacks occur even between Church members, involving 'medicine' blessed by the Church founder. Chief aim of baptism in 'Jordan' is to detect witches.

The prophets' diagnosing techniques resemble those of *nganga* and most Church members were 'made to join through illness'; they fear their illness will return if they turn renegade. Nevertheless, Churches experience a large turnover in membership because many people enter them only while they are in search of a particular aspect of 'power', and leave them again when the desire has been fulfilled. Even baptism is largely seen as a 'medicine'.

Mwari is presented to Church adherents as God of power whose chief gift is to enable one to 'live in strength'. However, for Church members the power emanating from the Church founder and the prophets is most important; it is conveyed on them through their contact with 'sanctified' symbols of power. Exorcism and cleansing rituals take an important place in the treatment of patients. Call-dreams are stereotyped as in the traditional Shona cosmology.

As I mentioned in the introduction, writers on the Shona do not seem to have found a generic term covering every form of ultra-human power. It may well be that such a term exists, but that it has not received the investigators' attention.[7] In spite of this, it is obvious that Shona regard the various ultra-human agencies which I have described, as subject to a single basic assumption about power. It is mirrored in the Shona proverb: 'Every power is subject to another power',[8] and in the many examples showing the inter-action of powers which I have presented, from the uneasy 'comrades-in-arms' relationship between chiefs and territorial spirits, to the unceasing war between the witch's 'medicines' and the diviner-healer's counter-'medicines'.

This last example is particularly interesting, because it shows that in this world-view, as Brain puts it, thief and policeman utilize the same power, but only the latter has the legitimate authority to do so.[9] The fact that the witch and the diviner-healer draw their powers from the same source has been seen to cast a shadow of ambiguity on the latter's status in public. However, ambiguity also taints the other wielders of power in the Shona 'spirit world'. One need only remember some of the myths about the first 'owners of the land' or the fact that stranger-spirits possess people for good as well as for ill. Also the ancestral-spirits' ambiguous nature has been extensively dealt with in chapter three.

In the ambiguity which the Shona attribute to their spirits they would seem to reflect the common experience that power has two faces. Depending on which side we find ourselves, we experience it as pleasant or unpleasant, enjoy it or suffer under it. The spirit concept, therefore, helps to explain on the one hand, the cause of unexpected misfortunes against which man is powerless. On the other hand, it allows for the fact that the clever will climb on the bandwagon of power and exploit, consciously or unconsciously, the widespread fear among the people of unpredictable powers. In this latter connection, one thinks particularly of the witch, the diviner-healer, and the various strategies which have been seen to be employed by those who, as mediums of territorial spirits and stranger-spirits, master the stereotype of 'possession'.

An important aspect of the conceptualization of spirits among the Shona, then, is that it attempts to account for all that is lying outside of man's ordinary and desirable experience, such as death, illness, anti-social inclinations and behaviour, sudden and unexpected luck or ill-fate, violent irruptions of nature, and the frustration of what would seem to be reasonable climatic expectations, etc. Any extraordinary or untoward experience gives rise to suspicions of interference by ultra-human powers. Thus, for instance, witchcraft has been seen to be, among other things, a rationalization of tension, and the witch as the stereotype of a person with evil intentions.

For the same reason, Shona spirits are always thought to be involved where individuals or communities find themselves in a situation of change: for example, territorial and ancestral spirits play a vital role during the succession of chiefs and heads of families; stranger-spirits may make a sudden appearance during a girls's difficult change from maidenhood to married life, or they may provide disillusioned housewives with the pretext for their having to join a 'club'; it will be remembered also that any host to stranger-spirits must undergo a *rite de passage* before he or she qualifies for his or her new status, for instance as diviner-healer. In this context one should recall the fact that a particularly large share of power is supposed to be won by those who fearlessly and unashamedly change their own situation through trespassing against the generally accepted norms.

It is also worthwhile to note the characteristics of the nature objects with which the Shona habitually associate their spirits. The wind's ubiquity images most aptly the supposed omnipresence of the spirits. Wind is also closely associated with the other two objects of nature figuring in the Shona spirit world: it causes sudden ripples to appear on the surface of pools and makes the leaves of trees rustle. Besides, water and trees are both, like the wind, characteristerized by their being subject to constant change.

Separation of subject and object

Like the Dinka,[10] the Shona have created numerous images or 'powers' as a consequence of whose action they can attribute their sufferings. However, not only suffering is thought to be caused by these 'powers'. According to Dinka accounts, the fetish *Mathiang Gok* 'works analogously to what, for [Whites], would be a guilty conscience'.[11] What *Mathiang Gok* is for the Dinka, namely 'a presence acting upon the self from without, and employed by some-one to do so', has its counterpart in various 'medicines' obtainable from Shona diviner-healers which are said to restore lost or stolen property to its rightful owner.[12] There exists also – for the same reason – a similarity between the supposed action of the *Mathiang Gok* and the Shona's avenging spirits. What Lienhardt has claimed in this connection with regard to the Dinka is also true of the Shona:

> The image . . . of the experience of guilty indebtedness . . . is extrapolated from the experiencing self. It comes (as memories often do) unwilled by the debtor, and is interpreted as a Power directed by the creditor.[13]

The earlier quoted observation by a Karanga, that 'only when a spirit hovers nearby will he watch us, and punish us when we do wrong',[14] can be interpreted in a similar manner: the 'spirit' – in this case a Shona man's ancestral spirit – is the objectification of a person's, or a number of persons', experiencing an uneasy conscience, resulting from the real or imagined neglect of those to whom respect and gratitude is owed. Murphree says of the Shona Apostles that in their view 'it is the passions of hatred and jealousy, which they con-consider to be characteristic of an undisciplined personality, that result in witchcraft'. As one informant told him,

> hatred and jealousy are the things that lead a person to witchcraft. For when a person thinks strongly on these things, the *shavi* appears that personifies them. Then this *shavi* goes to live in that person's calabash.[15]

Lienhardt has noted that the separation of a subject and an object in experience enables the Dinka not only to free themselves symbolically from what they must otherwise passively endure,[16] but provides them also with the possibility of creating a form of experience they desire. Lewis emphasized this possibility when he claimed that 'possession is essentially a philosophy of power'.[17] In chapters four and five, his analysis of possession and possession cults has helped me to throw much light on the concepts which underlie the Shona belief in the powers of stranger-spirits, witches, and diviner-healers.

The separation of a subject and an object in experience which the Shona share with the Dinka, through its elaboration in the concept of 'possession', provides them with an opportunity to display 'abnormal behaviour' in a society in which normality is the ideal, without incurring the usual negative sanctions. My speaking above of stereotypes – for example with regard to certain aspects of the *shavi* cults, particularly the readiness of Shona people to accommodate themselves to the spirits[18] – was intended to suggest that some of these strategies may actually come into operation by way of subconscious processes.

Traditional Shona 'power' and the 'Churches of the Spirit'

With the Shona, however, not only possession, but their whole traditional cosmology has been seen to be a 'philosophy of power'. In the light of this fact it is not surprising that the 'Churches of the Spirit' focus their attention particularly on those aspects of the Christian Gospel which fit in with this traditional Shona 'philosophy of power'. Typically, their preoccupation with the 'powers' of old is expressed in the Shona Apostles' claim that the Holy Spirit gives them the power to 'purify the country', and that for this reason they esteem the 'charism of exorcising evil spirits' as the greatest of all.[19] Table V clearly shows that there are numerous parallels between key figures in these Churches and the holders of key roles in the traditional Shona religion.

Fortunately, Dr Peter Fry's study of spirit mediumship amongst a community of Zezuru people living near Salisbury has become available before the publication of this book.[20] Although my discussion of territorial spirit mediumship is largely based on the interpretation of research among the Korekore, my findings and his with regard to this subject are virtually identical. An extremely interesting aspect of Fry's study is his description of how successful Zezuru spirit mediums may rapidly build up a vast clientele and a network of quasi-dependent mediums. These latter, after their training, remain in contact with their masters, although they set up their own 'practices' as specialists in divining and healing techniques, or as witch detectors.[21] The pattern of the successful spirit medium's sudden rise to fame, due to 'his capacity as a diviner, his ability to foster a wide circle of adherents, his ability as a showman, and a certain amount of luck'[22] shows obvious parallels with the career of some of the founders of 'Churches of the Spirit'. This is a further confirmation of parallels which I have drawn between these founders and spirit mediums. Bishop Mutendi, for example, has done the same as Kaguvi, who became the most influential spirit medium during the Shona Rebellion; i.e. 'he brought thousands of Shona into membership of a new society, the true believers in the M'lenga,[23] with their distinguishing symbols and obligations and their own promises of divine favour'.[24]

It would go beyond the scope of this book to present a comparison between the Shona spirits and the various 'Principalities and Powers' which are mentioned in the New Testament. Such a presentation will have to be the result of another serious research in its own right, as it will have to take into account a large body of scriptural studies available on this subject. The contribution of my study to a sound foundation for an emerging African Christian Theology consists in my having bared the roots of the traditional Shona cosmology. This cosmology has been found to provide the Shona with a systematic, all-embracing view of the world which needs no opening towards a transcendent, ultimate source of every power. It has also been seen that the 'Churches of the Spirit' are far from having confronted this traditional cosmology with the Gospel of Jesus Christ. Rather, they appear to be an example of what I have termed in my introduction the 'South Americanization' of the Church. Christ came to set men free from their bondage to 'the ruling spirits of the universe'.[25] Of his own free will, he took the nature of a servant, humbly walking the path of obedience to death on the cross.[26] In contrast with his example, the 'Churches of the Spirit' focus their attention on that power which Simon, the 'converted' Samaritan magician, attempted to buy from the Apostles.[27]

NOTES

1. Cf. above, p. 17.
2. Cf. Durkheim (1915), pp. 376-83.
3. Cohen (1969), p. 217. Cf. above, p. 17.
4. Cf. Daneel (1974), p. 116.
5. I would prefer to speak of the 'social order'.
6. *meta* (beyond, behind) *ta physika* (nature).
7. None of the present sources on Shona religion approaches the subject from the angle which is typical of Lienhardt's *Divinity And Experience*, i.e. attempting to investigate, in a consistent manner, the link between people's lives in their particular habitat and their religious assumptions. It is, however,

striking how often one encounters the word 'power' in all these sources, particularly in those which are anthropologically the most reliable ones.

8. Hamutyinei and Plangger (1974), p. 430.
9. Brain (1973), p. 125. He employs the metaphor with regard to the Shonas' neighbours, the Amandebele. According to Brain, their ancestral spirits (*amadhlozi*) are clearly associated with witchcraft through the root *loz-*, which is one variant of *loj-*, *loy-*, and *log-*, found in many Bantu languages with the meaning of witchcraft. Cf. *uroyi* = wizardry in Shona.
10. Lienhardt (1961), p. 190.
11. *Ibid.*, p. 150.
12. Cf. above, p. 123.
13. Lienhardt (1961), p. 150.
14. Cf. above, p. 69, n. 67.
15. Murphree (1969), p. 107.
16. Cf. above, p. 19.
17. Lewis (1971), p. 204.
18. Cf. above, p. 98.
19. Cf. above, pp. 151, 180.
20. Fry (1976).
21. *Op. cit.*, p. 38-42.
22. *Ibid.*, p. 41.
23. M'lenga is another form of spelling for Mulenga, the personal representative of Mwari, by whom Kaguvi claimed to be possessed.
24. Fry, *op cit.*, p. 50, quoting Ranger (1967), p. 225.
25. Col 2, 20.
26. Cf. Phil 2, 6-8.
27. Acts 8, 9-24.

BIBLIOGRAPHY

Abbot, W. M.: *The Documents of Vatican II.* London: Geoffrey Chapman, 1967.

Abraham, D. P.: The Monomotapa Dynasty. *Nada* 36:59–84, 1959.

The Role of 'Chaminuka' and the *Mhondoro* Cults in Shona Political History. In Stokes E. & Brown R. (eds.): 28–46, 1966.

Akers, M.: *Encyclopaedia Rhodesia.* Salisbury: The College Press, 1973.

Alpers, E.: The Mutapa and Malawi Political Systems to the Time of the Ngoni Invasions. In Ranger, T. O. (ed.): 1–28, 1968.

Aquina, M.: (N.B.: M. Aquina publishes also under the name A. K. H. Weinrich, q.v.) The People of the Spirit: An Independent Church in Rhodesia. *Africa* 37:203–18, 1967.

A Sociological Interpretation of Sorcery and Witchcraft Beliefs among the Karanga. *Nada* IX, 5:47–53, 1968.

Zionists in Rhodesia. *Africa* 39:113–36, 1969.

Spirit Beliefs among the Karanga. *Nada* X, 5:51–5, 1973.

Atkinson, N. D.: The Missionary Contribution to Early Education in Rhodesia. In Dachs, A. J. (ed.): 83–96, 1973.

Baden-Powell, R. S. S.: *The Matabele Campaign* 1896. London: Methuen & Co., 1897.

Banton, M. (ed.): *Anthropological Approaches to the Study of Religion.* London: Tavistock Publications, 1966.

Barrett, D. B.: *Schism and Renewal in Africa.* London: Oxford University Press, 1968.

(ed.): *African Initiatives in Religion.* Nairobi: East African Publishing House, 1971.

Baumann, H.: *Schöpfung und Urzeit des Menschen im Mythos der Afrikanischen Völker.* Berlin: Dietrich Reimer/Andrews & Steiner, 1936.

Beach, D. N.: The Initial Impact of Christianity on the Shona: The Protestants and the Southern Shona. In Dachs, A. J. (ed.): 25–40, 1973.

Beattie, J. H. M.: The Ghost Cult in Bunyoro. In Middleton, J. (ed.): 255–87, 1967.

Spirit Mediumship in Bunyoro. In Beattie, J. & Middleton, J. (eds.): 159–70, 1969.

& Middleton, J.: *Spirit Mediumship and Society in Africa.* London: Routledge & Kegan Paul, 1969.

Spirit Mediumship as Theatre. *RAIN* 20: 1–6, 1977.

Beidelman, T. O.: Kaguru Symbolic Classification. In Needham, R. (ed.): 128–66, 1973.

Bere-Chikara, F.: Cattle: The Life Blood of Shona Society. In Kileff C. & P. (eds.): 20–3, 1970.

Berlyn, Ph.: The Keeper of the Spirit of Nehoreka. *Nada* X, 4:55–9, 1972.

Bhebe, N. M. B.: Missionary Activity among the Ndebele and Kalanga: A Survey. In Dachs, A. J. (ed.): 41–52, 1973.

Bourdillon, M. F. C.: *Some Aspects of Shona Conceptualization.* B. Litt. Thesis: Oxford, 1968.

Some Aspects of the Religion of the Eastern Korekore. D. Phil. Thesis: Oxford, 1972a.

The Manipulation of Myth in a Tavara Chiefdom. *Africa* 42:112–21, 1972b.

Spirit Mediums in Shona Belief and Practice. *Nada* XI, 1:30–7, 1974.

The Shona Peoples: An Ethnography of the Contemporary Shona, With Special Reference to their Religion. Gwelo: Mambo Press. 1976.

Traditional Religion and an Independent Church. In Bourdillon, M. F. C. (ed.): 193–208, 1977.

(ed.): *Christianity South of the Zambezi II.* Gwelo: Mambo Press, 1977.

Brain, J. L.: Ancestors as Elders in Africa – Further Thoughts. *Africa* 43:122–33, 1973.

Breitenbach von, F.: *The Indigenous Trees of Southern Africa.* Vol. 2. Pretoria: The Government Printer, 1965.

Bullock, C.: *The Mashona.* Cape Town: Juta & Co., 1928.

The Mashona and the Matabele. Cape Town: Juta & Co., 1950.

Burke, E. E. (ed.): *The Journals of Carl Mauch: His Travels in the Transvaal and Rhodesia.* Salisbury: National Archives of Rhodesia, 1969.

Burridge, K. O. L.: *New Heaven, New Earth: A Study of Millenarian Activities.* Oxford: Blackwell, 1969.

Buxton, J.: *Religion and Healing in Mandari.* Oxford: Oxford University Press, 1973.

Cockcroft, I. G.: The Mlimo (Mwari) Cult. *Nada* X, 4:83–92, 1972.

Cohen, A.: Political Anthropology: The Analysis of the Symbolism of Power Relations. *Man* 4 (n.s.): 215–35, 1969.

Colson, E.: Spirit Possession among the Tonga of Zambia. In Beattie, J. & Middleton, J. (eds.): 69–103, 1969.

Territorial Cults of the Tonga of Zambia. Paper read at ASA Conference 1976, Manchester, 1976.

Crawford, J. R.: *Witchcraft and Sorcery in Rhodesia*. Oxford: Oxford University Press, 1967.

Dachs, A. J. (ed.): *Christianity South of the Zambezi I*. Gwelo: Mambo Press, 1973.

Daneel, M. L.: *The God of the Matopo Hills: An Essay on the Mwari Cult in Rhodesia*. The Hague: Mouton & Co., 1970.

Old and New in Southern Shona Independent Churches. Vol. I: *Background and Rise of the Major Movements*. The Hague/Paris: Mouton, 1971.

Old and New in Southern Shona Independent Churches. Vol. II: *Church Growth: Causative Factors and Recruitment Techniques*. The Hague/Paris: Mouton, 1974.

Davies, D. K.: *Race Relations in Rhodesia: A Survey for* 1972–73. London: Rex Collings, 1975.

Dos Santos, J.: *Ethiopia Oriental I*, bk. 1, cc. 2 & 4–15. Evora, 1609.

Douglas, M.: Animals in Lele Religious Symbolism. *Africa* 27:46–58, 1957.

(ed.): *Witchcraft Confessions and Accusations*. London: Tavistock, 1970.

Durkheim, E.: *The Elementary Forms of the Religious Life*. London: George Allen & Unwin Ltd., 1915.

Edwards, W.: From Birth to Death. *Nada* 7:16–42, 1929.

Evans-Pritchard, E. E.: *Witchcraft, Oracles and Magic among the Azande*. Oxford: Oxford University Press, 1937.

Nuer Spear Symbolism. In Needham, R. (ed.): 92–108, 1973.

Fortes, M. & Patterson, S. (eds.): *Studies in African Social Anthropology*. London: Academic Press, 1975.

Fortune, G.: *A Shona Nheketerwa*. In Fortes, M. & Patterson, S. (eds.): 207–19, 1975.

Fowle, J. L.: The Tribal Installation Ceremony of Chief Chirau. *Nada* X, 5:29–31. 1973.

Frobenius, L.: *Erythräa: Länder und Zeiten des Heiligen Königsmordes*. Berlin/Zürich: Atlantis-Verlag, [1930].

Fry, P.: *Spirits of Protest: Spirit-Mediums and the Articulation of Consensus among the Zezuru of Southern Rhodesia (Zimbabwe)*. Cambridge: Cambridge University Press, 1976.

Fynes-Clinton, S. H.: The History of the Msana Chieftainship. *Nada* X, 4: 60–3, 1972.

Garbett, G. K.: *Growth and Change in a Shona Ward. Occasional Paper No.* 1. Salisbury: University College of Rhodesia, Department of African Studies, 1960.

The Political System of a Central African Tribe with Particular Reference to the Role of Spirit Mediums. Ph.D. Thesis: Manchester, 1963.

Spirit Mediums as Mediators in Korekore Society. In Beattie, J. & Middleton, J. (eds.): 104–27, 1969.

Garlake, P. S.: *Great Zimbabwe*. London: Thames & Hudson, 1973.

Gelfand, M.: *Shona Ritual: With Special Reference to the Chaminuka Cult*. Cape Town: Juta & Co., 1959.

Shona Religion: With Special Reference to the Makorekore. Cape Town: Juta & Co., 1962.

Witch Doctor: Traditional Medicine Man of Rhodesia. London: Harvill Press, 1964.

The African Witch. Edinburgh: E. & S. Livingstone, 1967.

A Nganga who has Adopted 2 Faiths. *Nada* X, 3:73–6, 1971.

The Genuine Shona. Gwelo: Mambo Press, 1973.

Medicine and the Christian Missions in Rhodesia. In Dachs, A. J. (ed.) :109–24, 1973.

Gluckman, M.: *Rituals of Rebellion in South-East Africa*. Manchester: Manchester University Press, 1954.

Custom and Conflict in Africa. Oxford: Basil Blackwell, 1955.

Gomm, R.: Bargaining from Weakness: Spirit Possession on the South Kenya Coast. *Man* 10 (n.s.): 530–43, 1975.

Goody, J.: *Death, Property, and the Ancestors: A Study of the Mortuary Customs of the Lodagaa of West Africa.* London: Tavistock Publications, 1962.

Gray, R. F.: The Shetani Cult Among the Segeju of Tanzania. In Beattie, J. & Middleton, J. (eds.): 171–87, 1969.

Hammond-Tooke, W. D.: *The Bantu-Speaking Peoples of Southern Africa.* London: Routledge & Kegan Paul, 1974.

Hamutyinei, M. A. & Plangger, A. B.: *Tsumo-Shumo: Shona Proverbial Lore and Wisdom.* Gwelo: Mambo Press, 1974.

Hannan, M.: *Standard Shona Dictionary* (2nd ed.). Salisbury: Rhodesia Literature Bureau. 1974.

Hatendi, R. P.: Shona Marriage and the Christian Churches. In Dachs, A. J. (ed.): 135–49, 1973.

Hawkins, A M.: Black Labour and Employment in Rhodesia. *The South African Journal of African Affairs* 4: 8–23, 1974.

Heintze, B.: Der Südrhodesische *dziva*-Komplex. *Anthropos* 62: 337–68, 1967.

Hemans, T. J.: The Exorcism of Evil Spirits or Modern Exorcism. *Nada* X, 2: 64–6, 1970.

Holleman, J F.: *Shona Customary Law: With Reference to Kinship, Marriage, The Family, And the Estate.* Cape Town: Oxford University Press, 1952.

Accommodating the Spirit amongst some North-Eastern Shona Tribes. Rhodes-Livingstone Papers No. 22. London: Oxford University Press, 1953.

Chief, Council and Commissioner: Some Problems of Government in Rhodesia. Assen: Van Gorcum, 1968.

Horton, R.: The Kalabari World-View: An Outline and Interpretation. *Africa* 32:197–220, 1962.

African Traditional Thought and Western Science. *Africa* 37:50–71 & 155–87, 1967 a+b.

Types of Spirit Possession in Calabari Religion. In Beattie, J. & Middleton, J. (eds.): 14–49, 1969.

African Conversion, *Africa* 41:85–108, 1971.

On the Rationality of Conversion. *Africa* 45:219–35 & 373–98, 1975 a+b.

Houghton, D. H.: The Process of Economic Incorporation. In Hammond-Tooke, W. D. (ed.): 401–8, 1974.

Howman, R. H. G.: Chief and Councils in Southern Rhodesia. Apthorne, R. (ed.): *From Tribal Rule to Modern Government*. The Thirteenth Conference Proceedings of the Rhodes-Livingstone Institute for Social Research. Lusaka: 37–48, 1959.

Hughes, A. J. B.: Tribal Land Tenure – An Obstacle to Progress? Special Reference to the Tribal Trust Land of Rhodesia. *The South African Journal of African Affairs* 1:56–73, 1971.

Hugo, H. G.: The Mashona Spirits. *Nada* 13:52–8, 1935.

Johnston, T. F.: Tsonga Musical Performance in Cultural Perspective (South Africa). *Anthropos* 70: 761–99, 1975.

Jones, A. M.: African Music. *Rhodes-Livingstone Museum Occasional Papers* No. 21:16–19, 1943.

Jules-Rosette, B.: Song and Spirit: The Use of Songs in the Management of Ritual Contexts. *Africa* 45:150–65, 1975.

Katsukunya, V.: Makate and Nehoreka. *Nada* X, 3:77f. 1971.

Kay, G.: Changing Patterns of African Population in Zambia and Rhodesia. *The South African Journal of African Affairs*. 2:19–31, 1972.

Keesing, H.: Death, Property and the Ancestors: A Reconsideration of Goody's Concepts. Africa 40: 40–9, 1970.

Kenyatta, J.: *Facing Mount Kenya: The Tribal Life of The Gikuyu*. London: Secker and Warburg, 1953.

Kileff, C. & P. (eds.): *Shona Customs: Essays by African Writers*. Gwelo: Mambo Press, 1970.

Kopytoff, I.: Ancesters as Elders in Africa. *Africa* 41:129–42, 1971.

Kuper, H.: *An African Aristocracy: Rank among the Swazi*. Oxford: Oxford University Press, 1947.

Costume and Cosmology: The Animal Symbolism of the *Ncwala*. *Man* 8 (n.s.): 613–30, 1973.

Larson, T. J: The Spirit of the Ancestors and the *mandengure* Ceremony of the Hambukushu of Ngamiland. *Anthropos* 66:53–70, 1971.

Latham, C. J. K.: Munhumutapa: Oral Traditions. *Nada* X, 4:77–82, 1972.

The Social Organisation of the Mashona. *Nada* XI, 1:96–108, 1974.

Leach, E. R.: *Rethinking Anthropology*. London: The Athlone Press, 1961.

Lee, S. G.: Spirit Possession among the Zulu. In Beattie, J. & Middleton, J. (eds.): 128–56, 1969.

Lewis, I. M.: Spirit Possession in Northern Somaliland. In Beattie, J. & Middleton, J. (eds.): 188–217, 1969.

A Structural Approach to Witchcraft and Spirit-Possession. In Douglas, M. (ed.): 293–309, 1970.

Ecstatic Religion: An Anthropological Study of Spirit Possession and Shamanism. Pelican Anthropology Library: Penguin Books, 1971.

Lienhardt, G.: *Divinity and Experience: The Religion of the Dinka.* Oxford: Oxford University Press, 1961.

Littlejohn, J.: Temne Right and Left: An Essay on the Choreography of Everyday Life. In Needham. R. (ed.): pp. 288–98, 1973.

Loney, M.: *Rhodesia: White Racism and Imperial Response.* Penguin Books. 1975.

Magava, E. B.: African Customs Connected with the Burial of the Dead in Rhodesia. In Dachs, A. J. (ed.): 151–57, 1973.

Makamure, D. M.: Cattle and Social Status. In Kileff, C. & P. (eds.) 14–16, 1970.

Makarius, L.: The Magic of Transgression. *Anthropos* 69:536–52, 1974.

Marr, F.: Some Notes on Chief Sileya. *Nada* 39:81–4, 1962.

Marshall, L.: Kung Bushman Religious Beliefs. *Africa* 32:221–51, 1962.

Martin, M.-L.: *The Biblical Concept of Messianism and Messianism in South Africa.* Morija: Sesuto Book Depot, 1964.

The Mai Chaza Church in Rhodesia. In Barrett, D. B. (ed.): 109–21, 1971.

Mbee, G.: Letter from Mbugwe, Tanganyika. *Africa* 35:198–208, 1965.

Mbiti, J. S.: *African Religions and Philosphy.* London: Heinemann, 1969.

Middleton, J. (ed.): *Gods and Rituals: Readings in Religious Beliefs and Practices.* New York: The Natural History Press. 1967.

Spirit Possession Among the Lugbara. In Beattie, J. & Middleton, J. (eds.): 220–31, 1969.

Murdock, G. P.: *Africa: Its Peoples and their Culture History.* New York/London: McGraw-Hill, 1959.

Murphree, M. W.: *Aspects of Contemporary Shona Religion: With Special Reference to the Budjga.* Ph.D. Thesis: University of London, 1965.
Christianity and the Shona. London: The Athlone Press, 1969.

Mushoma: The Native Approach to Religion. *Nada* 10: 27–34, 1934.

Nada: *The Southern Rhodesian Native Affairs Department Annual.* 1923–1963.
The (Southern) Rhodesia Ministry of Internal Affairs Annual. Salisbury: The Government Printer, 1964.

Needham, R.: Percussion and Transition. *Man* 2 (n.s.): 606–14, 1967.
(ed.) *Right and Left: Essays on Dual Symbolic Classification.* Chicago: University of Chicago Press, 1973.

Nobbs, E. A.: *Guide to the Matopos.* Cape Town: Maskew Miller, 1924.

Nutting, A.: *Scramble for Africa: The Great Trek to the Boer War.* London: Constable. 1970.

Oosthuizen, G. C.: *Post-Christianity in Africa: A Theological and Anthropological Study.* Stellenbosch:T. Wever, 1968.

Pardy, A. A.: Notes on Indigenous Trees and Shrubs of Southern Rhodesia. *Rhodesian Agricultural Journal* 48: 264, 1951.
49:172, 1952.

Parkinson, D. K.: The Varozwi of Chief Murambgha. *Nada* X, 4:64–8, 1972.

Peaden W. R.: Zionist Churches in Southern Mashonaland 1924–1933. *The Bulletin of the Society for African Church History* 3:53–67, 1970.
Nenguwo Training Institution and the First Shona Teachers. In Dachs, A. J. (ed.): 71–82, 1973.

Pearce Commission: Report on the Commission on Rhodesian Opinion under the Chairmanship of the Right Honourable The Lord Pearce. London: Her Majesty's Stationery Office, 1972.

RAIN: *Royal Anthropological Institute News.* London.

Ranger, T. O.: The Role of Ndebele and Shona Religious Authorities in the Rebellions of 1896 and 1897. In Stokes, E. & Brown, R. (eds.): 94–136, 1966.

Revolt in Southern Rhodesia 1896–7: *A Study in African Resistance*. London: Heinemann, 1967.

(ed.): *Aspects of Central African History*. London: Heinemann, 1968.

The African Voice in Southern Rhodesia, 1898–1930. London: Heinemann, 1970.

& Kimambo, I. N. (eds.) *The Historical Study of African Religion: With Special Reference to East and Central Africa*. London: Heinemann, 1972.

& Weller, J. (eds.): *Themes in the Christian History of Central Africa*. London: Heinemann, 1975.

Dance and Society in Eastern Africa 1890–1970: *The Beni Ngoma*. London: Heinemann, 1975a.

The Historiography of the Mwari Cult of Rhodesia: Origins and Early History. Paper Read at ASA Conference 1976, Manchester, 1976.

Redmayne, A.: Chikanga:An African Diviner with an International Reputation. In Douglas, M. (ed.): 103–27, 1970.

Rhodesia: *The Demand for Independence in Rhodesia: The Dombashawa 'Indaba'*. Salisbury: The Government Printer, 1964.

Robinson, K. R.: A Diviner from Malawi. *Nada* X, 3:11–12, 1971.

Rolin, H.: *Les Lois et L'Administration de la Rhodesie*. Bruxelles: Emile Bruylant, 1913.

Rungano cf. Daneel (1971).

Schebesta, P.: *Portugals Konquistamission in Südost-Afrika: Missionsgeschichte Sambesiens und des Monomotapareiches* (1560 - 1920). St. Augustin: Steyler Verlag, 1966.

Schoffeleers, M.: The Interaction of the M'Bona Cult and Christianity, 1859–1963. In Ranger, T. O. & Weller, J. (eds.): 14–29, 1975.

Shack, W. A.: Hunger, Anxiety, and Ritual: Deprivation and Spirit Possession among the Gurage of Ethiopia. *Man* 6 (n.s.): 30–43, 1971.

Shorter, A.: The *migawo:* Peripheral Spirit Possession and Christian Prejudice. *Anthropos* 65:110–26, 1970.

Sibisi, H.: The Place of Spirit Possession in Zulu Cosmology. In Whisson, M. G. & West, M. (eds.): 48–57, 1975.

Simon, H. J.: *Chief, Council and Commissioner* (Holleman, J. F., q.v.) Review Article. *African Social Research* 11:55–65, 1971.

Southall, A.: Spirit Possession and Mediumship among the Alur. In Beattie, J. & Middleton, J. (eds.): 232–72, 1969.

The Southern Cross: A Catholic Weekly Paper. Cape Town.

The Star: A daily newspaper. Johannesburg.

Stayt, H. A.: *The Bavenda.* Oxford: Oxford University Press, 1931.

Stokes, E. & Brown, R. (eds.): *The Zambesian Past: Studies in Central African History.* Manchester: Manchester University Press, 1966.

Sturtevant, W. C.: Categories, Percussion and Physiology. *Man* 3 (n.s.): 133f, 1968.

Sundkler, B. G. M.: *Bantu Prophets of South Africa.* London: Oxford University Press (2nd edition), 1961.

Tanner, R. E. S.: The Theory and Practice of Sukuma Spirit Mediumship. In Beattie, J. & Middleton, J. (eds.): 273–89, 1969.

Theal, G. McCall: *History and Ethnography of Africa South of the Zambezi.* Vol. I: The Portuguese in South Africa from 1505 to 1700. London: Swan Sonnenschein, 1907.

Torrend, J.: *A Comparative Grammar of the South African Bantu Languages.* London: Routledge and Kegan Paul, 1921.

Turner, V. W.: Colour Classification in Ndembu Ritual. In Banton, M. (ed.): 47–84, 1966.

Vambe, L.: *An Ill-Fated People: Zimbabwe Before and After Rhodes.* African Writers Series. London: Heinemann, 1972.

Van Warmelo, N. J.: The Classification of Cultural Groups. In Hammond-Tooke, W. D., (ed.): 56–84, 1974.

Verger, P.: Trance and Convention in Nago-Yoruba Spirit Mediumship. In Beattie, J. & Middleton, J. (eds.): 50–66, 1969.

Walker, S. S.: *Ceremonial Spirit Possession in Africa and Afro-America.* Leiden: Brill, 1972.

Watson, L.: *Supernature: A Natural History of the Supernatural.* London: Coronet (Hodder Paperbacks), 1974.

Weinrich, A. K. H.: *Modern Chieftainship and the Struggle for Power among the Karanga: A Comparison of the Political Process in Two Rhodesian Reserves.* Ph.D. Thesis: Manchester, 1965.

Chiefs and Councils in Rhodesia: Transition from Patriarchal to Bureaucratic Power. London: Heinemann, 1971.

An Aspect of Development of the Religious Life in Rhodesia. In Ranger, T. O. & Weller, J. (eds.): 218–37, 1975.

Welch, S. R.: *Portuguese Rule and Spanish Crown in Southern Africa* 1581–1640. Cape Town/Johannesburg: Juta & Co., 1950.

Werbner, R. P.: Atonement Ritual and Guardian-Spirit Possession among the Kalanga. *Africa* 34:206–22, 1964.

Whisson, M. G. & West, M. (eds.): *Religion and Social Change in Southern Africa.* Cape Town: David Philip, 1975.

Willis, R. G.: Pollution and Paradigms. *Man* 7:369–86, 1972.

Wilson, B. R. (ed.): *Rationality.* Oxford: Blackwell, 1975.

Wilson M.: *Religion and the Transformation of Society: A Study in Social Change in Africa.* Cambridge: University Press, 1971.

Young, J.: The Legendary History of the Hodi and Gorima Chiefs. *Nada* X, 2:49–60, 1970.

Zvogbo, C. J. M.: The Influence of the Wesleyan Methodist Missions in Southern Rhodesia. In Dachs, A. J. (ed.): 63–70, 1973.

ACKNOWLEDGEMENTS

Grateful acknowledgement is made to the following for permission to quote material in copyright:

Atlantis Verlag for extracts from *Erythäa*: *Länder Und Zeiten Des Heilgen Köñigsmordes* (1930) by L. Frobenius.

Dr. G. K. Garbett for extracts from 'The Political System of a Central African Tribe with Particular Reference to the Role of Spirit Mediums', Ph. D. Thesis, Manchester University (1963).

The International African Institute for excerpts from the following articles from their journal *Africa:*
'The People of the Spirit: An Independent Church in Rhodesia' (37, 203-218, 1967) by M. Aquina;
'African Conversion' (41,85-108, 1971) and 'On the Rationality of Conversion' (45, 219-235 & 373-398, 1975 a & b) both by R. Horton;
'Song and Spirit; The Use of Songs in the Management of Ritual Contexts, (45, 150-165, 1975) by B. Jules Rosette·

Mambo Press for extracts from *The Shona Peoples: An Ethnography of The Contemporary Shona, with Special Reference to Their Religion* (1976) by M. F. C. Bourdillon.

Mouton and Co. Publishers (The Hague) for extracts from *Old and New in Southern Shona Independent Churches* (Vol 1 – *Background and Rise of the Major Movements* and Vol 2 – *Church Growth: Causative Factors and Recruitment Techniques*) by M. L. Daneel.

Nada (The Annual of the Zimbabwe Ministry of Home Affairs) for an extract from 'Spirit Beliefs Among the Karanga' (x, 5: 51-55, 1973) by M. Aquina.

Northwestern University Press, Heinemann Educational Books Inc. and Terence Ranger for extracts from *The African Voice in Southern Rhodesia 1898-1930* (1970) by Terence Ranger.

Terence Ranger and Heinemann Educational Books Inc. for an extract from *Revolt in Southern Rhodesia 1896-7: A Study in African Resistance* (1967) by Terence Ranger.

The Royal Anthropological Institute of Great Britain and Ireland for an extract from 'Categories, Percussion and Physiology' by W. C. Sturtevant in *Man*, 3 (N.S.) 133f, 1968.

A. K. H. Weinrich for an excerpt from 'Modern Chieftainship and the Struggle for Power among the Karanga: A Comparison of of the Political Process in Two Rhodesian Reserves', Ph. D. Thesis, Manchester University, 1965.

INDEX

This index is arranged word-by-word. Specific entries are used, usually in direct form, *e.g.* 'founder-ancestors' not 'ancestors, founder'. Prepositions, conjunctions and articles at the beginning of subheadings have been ignored in determining the alphabetical order of subheadings.

Abbreviations

AAC	African Apostolic Church
BSA	British South Africa Company
IC	Independent Churches
UDI	Unilateral Declaration of Independence
ZAC	Zion Apostolic Church
ZAFM	Zionist Apostolic Faith Mission
ZANU	Zimbabwe African National Union
ZAPU	Zimbabwe African People's Union
ZCC	Zion Christian Church

passim following page references, *e.g.* 1–7 *passim*, means: references scattered ou the pages indicated.